Advanced Numerical Simulation Methods

Advanced Numerical Simulation Methods

From CAD Data Directly to Simulation Results

Gernot Beer

Institute for Structural Analysis, Graz University of Technology, Graz, Austria

CRC Press
Taylor & Francis Group
Boca Raton London New York

CRC Press is an imprint of the
Taylor & Francis Group, an **informa** business

A BALKEMA BOOK

The cover is a display of the CAD model and the simulation results (displacement contours) for two tunnels with a cross passage (copyright G. Beer).
Cover design by Gisela Beer, BSc (hons) in visual communication (Monash University).

CRC Press
Taylor & Francis Group
6000 Broken Sound Parkway NW, Suite 300
Boca Raton, FL 33487-2742

First issued in paperback 2020

© 2015 by Taylor & Francis Group, LLC
CRC Press is an imprint of Taylor & Francis Group, an Informa business

Typeset by MPS Limited, Chennai, India

No claim to original U.S. Government works

ISBN-13: 978-1-138-02634-6 (hbk)
ISBN-13: 978-0-367-78343-3 (pbk)

Library of Congress Cataloging-in-Publication Data

Beer, G. (Gernot)
  Advanced numerical simulation methods : from CAD data directly to simulation results / Gernot Beer.
      pages cm
  Includes bibliographical references and index.
  ISBN 978-1-138-02634-6 (hardcover : alk. paper) – ISBN 978-1-315-76631-7 (ebook : alk. paper)   1. Computer simulation.   I. Title.

  QA76.9.C65B44   2015
  003'.3–dc23
                                          2015023012

Table of contents

Preface

The real voyage of discovery consists not in seeking new landscapes, but in having new eyes.

M. Proust

In today's age one cannot imagine a world without numerical simulation. It plays an important role in engineering design, weather forecasting and many other aspects of our life. The design of a new type of aircraft such as the A380 for example would not have been possible without it. Simulation methods have become extremely sophisticated and we can now solve large fluid structure interaction problems involving millions or even billions of unknowns. This has been made possible by the tremendous increase in computing power, unthinkable a decade ago and the development of very sophisticated mesh generation programs.

However, as anyone that has been involved in serious simulation can attest, one of the main bottlenecks is still the need to generate a mesh for the analysis. Engineers use Computer Aided Design (CAD) software for the design, be it a new type of aircraft or underground works to create a digital representation of the geometry. The mesh generation process not only repeats this work, but also introduces an unnecessary approximation of the CAD representation. This is a far from ideal situation and came about because CAD and simulation software developed completely independently. The main aim of CAD software is a visual representation of the prototype whereas the aim of the simulation software is the prediction of its behavior, two completely different objectives. Therefore, the digital representation produced by the CAD programs is not *analysis suitable* and hence the need for mesh generation.

How I became interested in isogeometric analysis

I first became aware of the efforts by Tom Hughes and his group to address this problem, fairly late in 2011, after listening to a keynote lecture in the wonderful Greek island of Kos. I subsequently read the fine and very readable book "Isogeometric Analysis".

I soon became very excited about this development, as in my view it offered a real breakthrough in technology similar to the emergence of the Finite Element Method (FEM) or Boundary Element Method (BEM). A seamless integration of CAD and simulation has the potential of changing the industry.

After becoming emeritus professor and having enough time on my hands, I became fascinated by Non-Uniform Rational B-Splines (NURBS) and the possibilities they offered. I found that the MATLAB programming language (or its freeware counterpart OCTAVE) offered a quick and easy way to try out different things, the main advantage being the ease of generating graphical output.

As most engineers, I rely on a graphical representation to help me understand theoretical concepts. After understanding NURBS and their power, I became convinced that a seamless integration of CAD and simulation was a possibility.

The first task was to study the IGES standard for CAD data exchange, a 650 page document. I then developed a parser that could read and interpret the data from the ASCII file created by CAD programs. It turned out that CAD programs use trimming of NURBS surfaces, so there was the task of converting surface and trimming information into an analysis suitable form. This lead to the development of a simple algorithm which was published in the journal Computer Methods in Applied Mechanics and Engineering in 2015.

Next, I began to develop Finite Element software to understand the subtle differences in the implementation and the advantages that NURBS offered. It became soon clear that the NURBS functions were superior to Serendipity and Lagrange functions currently in use. I also realized that the isoparametric concept (using the same functions for the description of the geometry as for the approximation of the unknown) was not a very efficient way to proceed. Since the geometrical description is taken directly from the CAD program there is no need to refine it, but in most cases there would be a need to refine the approximation. Therefore, the isoparametric concept was abandoned.

Turning to 3-D solid analysis, it soon became obvious that mesh generation could only be avoided if the Boundary Element Method (BEM) was used since this method only involves a definition of bounding surfaces, the same as for CAD. Therefore, the emphasis in the further development was in the BEM.

In May 2013 Stephane Bordas and I organized a course on "Isogeometric methods for numerical simulation" at the International Center for Mechanical Sciences (CISM) in Udine, which was well attended. A research project funded by the Austrian Science Fund (FWF) enabled a small but very active group to be established at the Graz University of Technology, that made some significant progress especially in the field of the isogeometric BEM. Some of this work is included in this book, but at the time of publication a PhD thesis on "Seamless integration of CAD and simulation" is still in preparation.

Why this book was written

The idea of writing this book came from the desire to make a contribution towards the goal of achieving a seamless integration of CAD and simulation without mesh generation. This includes developing a toolkit that can be used by researchers as a basis from which to start new developments. The main emphasis of the book is on implementation and for each stage programs have been developed so readers can try out and get a feel for the new developments. The main aim is to introduce readers to the novel aspects offered by isogeometric modeling. It is not intended as a comprehensive treatise on simulation, so many advanced topics, such as geometrically nonlinear analysis and fluid structure interaction are left out.

This book is written by an engineer for engineers. It is hoped that mathematicians will excuse my sometimes liberal approach to mathematical theory and that the book will provide an impetus for the development of next generation simulation software that integrates seamlessly with CAD.

For whom this book was written

The book was written for users of simulation software, so they can understand the benefits of this new technology and demand progress from a somewhat conservative industry. It is written for software developers, so they can see that this is a technology with a big future. Finally, it is written for researchers with the hope that it will attract more people to work in this exciting new field.

How to read this book

The book is written like a road book, leading the reader on a journey towards understanding isogeometric analysis and the state of development. I have divided the book into stages and after each stage the reader will have gained knowledge that is required for the next stage. A road map is shown in the Introduction.

Programs available

OCTAVE functions, that have been used in this book, are available for readers. They should run with no or minor modifications in MATLAB. Send an e-mail to **gernot.beer@tugraz.at** with the subject **Book programs** and stating your name and affiliation to request access to a dropbox account that contains the sources. The NURBS toolbox, used by the software, can be downloaded free of charge from http://octave.sourceforge.net/nurbs/index.html.

Acknowledgements

I would like to acknowledge various people that have helped in the development of this book. Firstly, Thomas J.R. Hughes, who started this exciting new development of isogeometric analysis. Thanks are due also to my co-wokers Jürgen Zechner and Benjamin Marussig for their outstanding research work in the project "Fast isogeometric BEM" and to the Austrian Science Fund (FWF) for funding the research.

I would like to thank my colleagues Thomas-Peter Fries, Luiz Wrobel, Adrian Cisilino, Andre Pereira and Christian Duenser for reviewing some chapters and making suggestions for improvements. Thanks to Stephane Bordas, who was the co-organizer of the course "Isogeometric methods for numerical simulation" presented at the International Center for Mechanical Sciences (CISM), which gave some impetus to write this book. A special thank you to the developers of the NURBS toolkit and especially Rafael Vazquez for his support.

To my daughter Gisela, for designing the cover and the road map and to my wife Sylvia for carefully proofreading the manuscript. Last but not least thanks to the people at CRC Press for their support and excellent work in publishing.

Gernot Beer
Nelson Bay, Australia 2015

About the author

The author started numerical simulation as part of his PhD work in 1973 where he developed a simulation method for arc welding of steel plates. Since that time he has been active in modelling, mainly in the area of underground works (mining and tunnelling) and developed a commercial computer package (BEFE) that combines two methods of analysis. He has been involved as a consultant in many interesting projects around the world such as the design of caverns for the Hadron collider at CERN and an underground power station in Iran.

He has written 3 textbooks on the subject, the first one being about 2 methods (Finite Element and Boundary Element Method) and has coordinated many research projects, including the world's largest project on underground construction (EC project TUNCONSTRUCT).

Currently he is emeritus professor of Graz University of Technology, Austria and conjoint professor of the University of Newcastle, Australia.

Chapter 1

Introduction

I am a little world made cunningly of elements

Donne

For nearly all physical processes in nature a differential equation can be set up, using the fundamental laws of physics such as equilibrium, the preservation of energy and others. This, combined with laws describing the material behavior, allows setting up equations that can be solved analytically for a very limited class of problems.

Numerical simulation evolved from the need to solve real life problems, where exact solutions are not possible. Such solutions were required, for example, for the safe design or for prediction of behavior. Without numerical simulation we would be unable to design tall buildings or the next generation of aircraft. Numerical simulation always involves an approximation of the real world since most problems are too complex to be analyzed and need to be abstracted. For example, the A380 aircraft has millions of parts and it would be impossible to model it in all detail. Abstraction or simplification of the problem is one of the challenges of numerical simulation, that can not be taken over by a computer, at least in the forseeable future. However, as we will see next many serious mistakes can be made here. Even if the abstraction is handled correctly, there is still another aspect where errors can be made, namely in the approximation of the geometry and the known and unknown values. We will discuss the emergence of numerical simulation and the milestones associated with it next, before introducing the contents of the book.

1 A BRIEF HISTORY OF SIMULATION

1.1 The world's first simulation

Numerical simulation actually started quite early and was driven by the need to understand and predict behavior. An early example of simulation dates back to 1745 (see [4]) when Pope Benedict XVII was worried about the stability of the cupola of the St. Peter's dome since it was observed that it had developed cracks. The cupola had circumferential iron rings installed to ensure stability, but it was questioned if these rings were adequate. The dome was built by artists with knowledge handed down by generations, but no design in the modern sense was done.

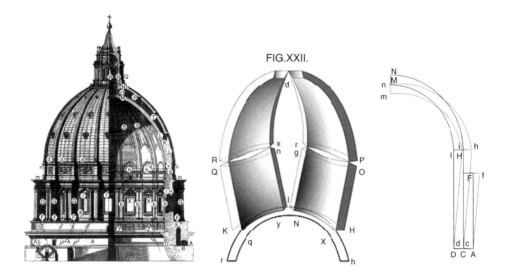

Figure 1 Cupola of St. Peter's dome and model abstractions.

The pope employed 3 mathematicians to solve the problem. The first task for them was to simplify the problem. This involved the following steps:

- Understand the basic mechanics of the problem
- Based on this understanding, identify the important mechanisms
- Develop a mechanical model

The mathematicians realized, that the main driving force for the development of the cracks was the horizontal trust generated by the weight of the cupola. If they could determine the horizontal thrust, then they could also determine the circumferential force and find out if the iron rings were adequate. However, the mathematical tools available at the time were very limited. For example, they could not deal yet with deformable bodies of arbitrary shape and therefore the mechanical model they devised, involved rigid bodies. To be able to compute the horizontal thrust, hinges had to be assumed as shown in Figure 1. As simulating a 3-D problem was also beyond their capabilities, they simplified it to a plane problem as shown on the right of Figure 1.

However, there was at this stage a good understanding of equilibrium and methods for determining it, published by Johann Bernoulli 27 years earlier: The principle of virtual work or, in particular, the principle of virtual displacements, which states that for a system in equilibrium the work done by the forces times virtual displacements should be zero. It is interesting that this principle is the one used in modern numerical methods, as will be demonstrated later.

The idea of the mathematicians was to apply a virtual displacement to the hinged structure so the horizontal thrust does virtual work. The other virtual work done is due to the derived displacements of parts of the structure times the gravitational force due to self weight. The equation for the equilibrium was quite simple:

$$H \cdot \delta u + \sum \delta W = 0 \tag{1}$$

Figure 2 The Pope and two of the mathematicians employed by him.

where H is the horizontal thrust, δu is the virtual displacement and $\sum \delta W$ is the sum of virtual work done by the gravitational forces. The document produced by the mathematicians was quite voluminous and the main part dealt with the detailed examination of the cracks and the determination of gravitational forces.

The conclusion of the document, however, was quite wrong. The horizontal thrust computed was so large that the dome would have collapsed. The reason for this was the abstraction, which was too extreme and did not capture the mechanics of the problem, in particular the 3-D effects. This is an extreme example where the abstraction caused the results of the simulation to be meaningless. There are quite a few examples of wrong results being obtained even in modern numerical modeling, one of which will be shown later.

Figure 3 shows a re-analysis of the dome with the Finite Element Method. Because of cyclic symmetry only 1/8 of the dome needed to be analyzed (for the case of gravity loading) with the appropriate boundary conditions, without any effect on the results. Such clever model reduction should always be applied as it can save a lot of modeling effort and simplify the analysis.

The crack patterns predicted by the model matched the observed ones well and the horizontal thrust and the circumferential force computed was of a magnitude that could be carried by the iron rings. Therefore, the conclusion of this analysis was that the dome was stable and as can be observed by any visitor to Rome, the St. Peters cathedral still stands and is likely to do so for many centuries.

1.2 Emergence of mathematics and mechanics

Although the first attempt at numerical modeling was not successful, it was an important milestone in the recognition of the importance of mathematics and mechanics. Various numerical methods evolved in the early 1900s to solve differential equations. Notably the methods by Ritz published in 1909, Trefftz in 1927 and the emergence of the finite difference method in 1930. The original idea of Ritz was to approximate the

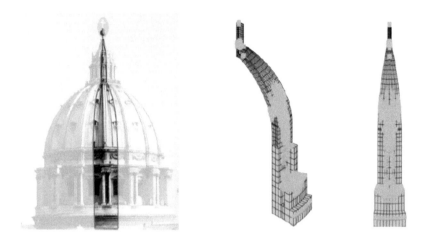

Figure 3   Finite Element analysis of St. Peter's dome and resulting crack patterns.

Figure 4   O. C. Zienkiewicz.

solution by functions multiplied with unknown parameters, the magnitude of which can be determined by the virtual work principle that had been used by the mathematicians in the first simulation. The idea by Trefftz was to use fundamental solutions of the differential equations and approximate the boundary values only. Finally the idea of the finite difference method is to numerically approximate the differentiation. All of the methods lead to simultaneous equations, the solution of which was too cumbersome or impossible without a digital computer.

1.3   Computer age

Only when the first computers became available around 1960 did these methods take off. First the finite difference method and then, based on the ideas of Ritz, the Finite Element method.

Based on the idea of Trefftz Boundary Element methods were developed at about the same time.

The Finite Element Method (FEM)   This method is based on an original idea by Ritz, who proposed that the variation of the unknown is approximated by functions that are continuous in the domain. First applications of the method were restricted to cases where such continuous functions could be found, for example for circular plates. It soon became obvious that this restriction had to be lifted for the method to be useful in practice. In the early 1960s, when the first digital computers became available, the idea was proposed to subdivide the domain and to use functions which were continuous over the subdomain only. The idea of the Finite Element method was born. Due to the initiative of O. Zienkiewicz the method has really taken off. Soon very powerful programs were developed first by NASA (NASTRAN) and then by others. As computers became more and more powerful the capabilities of the programs increased and soon the degree of abstraction of the problem could be reduced considerable.

A good example, that abstraction is still important and if not done right can lead to wrong results is the collapse of the Sleipner oil platform (see [3]). A catastrophic failure of the platform occurred in 1991 and was later attributed to a faulty design based on results of a numerical simulation, which underestimated the stress by 45%. A sketch of the platform is shown in Figure 5. Basically the platform is constructed floating, with the caissons providing buoyancy.

After the construction some air is removed from the caissons, so that the platform slowly descends to the sea floor. During this process the failure occurred because the design of the reinforcement was based on results of the numerical simulation which were in error. The 3D mesh used for the analysis is shown in Figure 5 (top right). This mesh simulates the whole caisson structure and the only abstraction made was to assume two planes of symmetry. In contrast to the world's first simulation the abstraction appears to be much less and gives the impression that the total structure has been modeled accurately.

While this is true, another error associated with the approximation of the unknown has occurred. To limit the size of the system of equations to be solved, solid Finite Elements with a linear approximation of the displacements were used with only one element across the thickness. Since in the FEM the strains and subsequently the stresses are computed by taking a derivative of functions approximating the displacements, the prediction of stress was not very accurate.

To check the results, a detailed plane analysis was done after the event of a critical horizontal section through the cells, a detail of which is shown in Figure 5 (bottom). The failure occurred in a part of a cell that was above the internal water level and it is attributed to the external water pressure inside the triangular section where 3 caissons meet (referred as tri-cell). Figure 6 shows the distribution of the maximum (tensile) principal stress for the original and a refined mesh. It can be seen that the stress prediction of the original mesh is very poor. Especially in the area where the failure occurred the finer mesh predicts much higher tensile stress than the original coarse mesh.

This is a good counter-example of the previous example, where the degree of abstraction was too extreme. Here the degree of abstraction was actually quite low. It is a good example, however, of another aspect that needs to be considered in numerical simulation, namely how good the approximation of the unknown is. At the time of the analysis there was an euphoria and the belief that numerical methods were so powerful that very little abstraction was necessary. As the case shows, this is a fallacy. Even

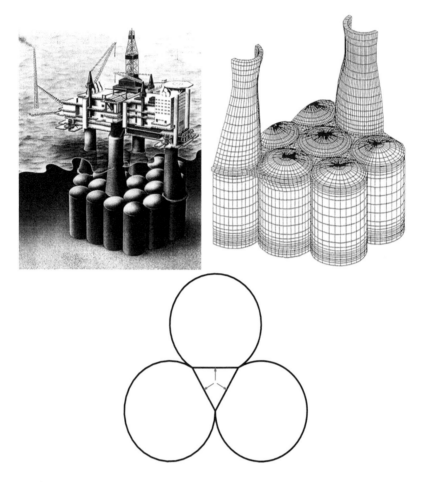

Figure 5 The Sleipner oil platform: Schematic, mesh used for analysis and detail of horizontal section through cells with applied loading on the tri-cell.

with the powerful programs available today, an intelligent abstraction is necessary. In particular, results of numerical simulation should always be treated with suspicion. Simple, "back of the envelope" type of calculations should always be made. This means that abstraction is still a very important aspect of simulation.

The Boundary Element Method (BEM) The original idea of Trefftz to use fundamental solutions of the governing differential equations led to the development first of the Boundary Integral Equation method and then of the Boundary Element method. The development of the method was quite different from its "big sister" the FEM. The reason for this is that there was much higher demand on mathematical skills (especially when dealing with singular integrals) and in the early days the method was restricted

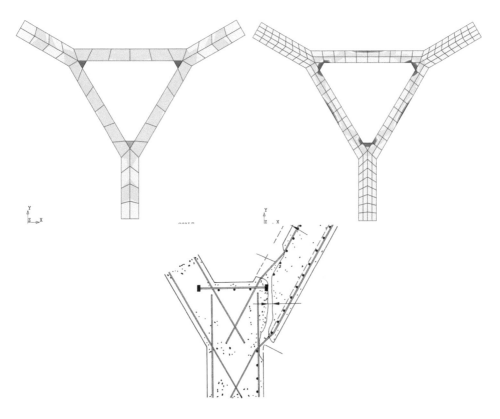

Figure 6 Results (contours of maximum principal stress) of a detailed plane analysis of the part of the cross-section where the cells meet (tri-cell). Left: original mesh; Right: refined mesh; Below: Sketch of the failure.

to homogeneous domains with linear (elastic) material behavior. Also, early books on the method were not so easy to understand for engineers.

The main attraction of the method is that only boundaries of the problem need to be defined. It is therefore an ideal companion to CAD and the only method where it is possible to avoid mesh generation. The approximation of the unknown occurs only on the boundary and inside the domain functions are used that satisfy the governing differential equation exactly. This means that in elasticity the solutions inside the domain satisfy equilibrium and compatibility exactly. Primary results are obtained on the boundary and interior results are computed by post-processing as will be explained in detail later. The method is therefore ideal for predicting stress concentrations on the boundary. As an example we show a reanalysis of the Sleipner tri-cell in Figure 7 where contours of stress are plotted on a user defined result plane. Another beneficial aspect of the method is that the user determines where the results should be shown (in contrast to the FEM where results are computed everywhere).

The main attraction of the method, however is, that it can deal with infinite domains without the need for truncation. The early applications of the method

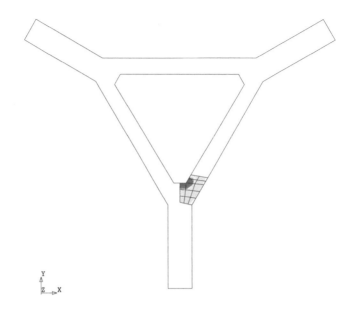

Figure 7 Boundary Element analysis of Sleipner tri-cell with contours of maximum principal stress plotted on a result plane.

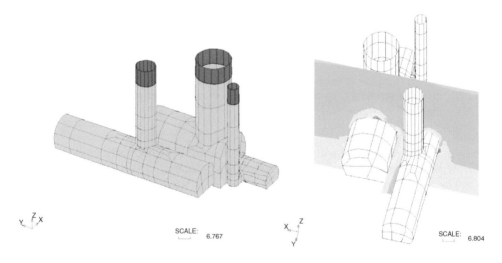

Figure 8 Boundary Element mesh of CERN caverns and minimum (compressive) principal stress plotted on a result plane.

were therefore in geomechanics in particular mining and in underground excavations (see for example [2]). Figure 8 shows an application of the BEM in geomechanics, namely the analysis of the upgrade of caverns at the Large Hadron Collider of CERN (see [1]).

Coupled method To overcome the fact that the BEM could not deal with heterogeneous domains and non-linear material behavior, a coupling with the FEM was suggested. In his famous paper [5] Ziekiewicz proposed a marriage of convenience between the two methods, where the FEM was used in areas where material nonlinear behavior was expected to occur. The coupled method has not really taken off, mainly because of the complexity in the mesh generation.

2 BASIC STEPS IN SIMULATION

The basic steps in simulation are:

- Approximation of the geometry of the problem
- Approximation of the field variables
- Solution
- Recovery of results

2.1 Geometry description

The first task is to approximate the geometry of the problem to be analyzed. For 3-D solids this is done by describing its bounding surfaces. For an approximation of the real world by a plane analysis the object is described by bounding lines. In some cases the domain to be analyzed is much larger than the area of interest is truncated for the simulation at a distance, chosen in such a way that its influence is negligible. An extreme case is found in the simulation of geotechnical problems where the domain (the earth's crust) can be assumed to be of infinite extent.

In the FEM the usual way to describe the geometry of the problem is by approximation with Finite Elements. Finite Elements are simple building blocks that are connected together to model complicated geometrical shapes. In 3-D these are solid elements of prismatic or tetrahedral shape, whose shape is determined by interpolation between fixed points (nodal points). Interpolation functions used by most commercial computer programs are Seredipity functions or Lagrange polynomials of order 1 or 2 (linear or quadratic).

This process, where the geometry is described by discrete points and interpolation functions is known as *discretization*. The assembly of finite building blocks is called a mesh and the approximation of the real geometry is achieved by mesh generation. In the BEM mesh generation is made much easier because only the bounding surfaces need to be discretized into Boundary Elements.

Mesh generation usually starts by defining the boundary of the domain to be analyzed. This can be achieved via one of the many available mesh generation programs. Boundaries can be defined by using direct interactive input with a graphics toolkit consisting of lines, arcs, splines etc. Alternatively, many mesh generation programs offer the possibility of reading a file output from a Computer Aided Design (CAD) program.

Since CAD is nearly always used for the design, this is an obvious approach. Most CAD programs are able to generate text files in a standard format that can be read

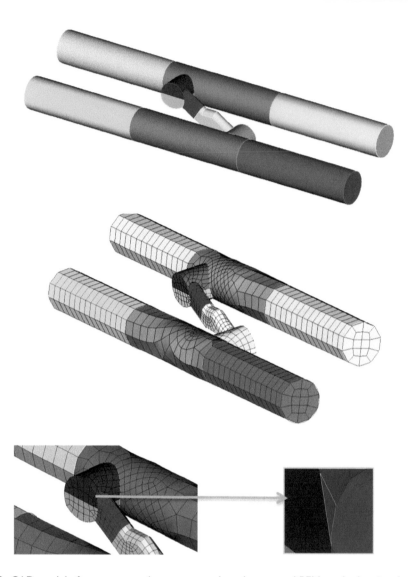

Figure 9 CAD model of cross passage between tunnels and generated BEM mesh, showing detail and an element with bad aspect ratio.

by mesh generators. The impression may be gained that this process is completely automatic, but this is not the case.

As an example we show the generation of a BEM mesh using output from a CAD program. Figure 9 shows the CAD model of tunnels at different levels and a cross passage between them[1]. A BEM mesh was generated using the freeware program CUBIT with linear quadrilateral elements.

---

[1]This is based on an actual project that unfortunately was not realized because of budget cuts.

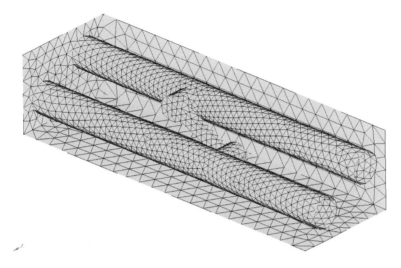

Figure 10 Finite Element Mesh of tunnel problem (only some surfaces of elements are plotted).

It can be seen that even with a fair number of elements, the geometry is not approximated very well. In addition, elements are generated with a bad aspect ratio, i.e. where the angle between vertices is greater than 180° or less than 0°. In this case, a non-unique mapping occurs, resulting in a zero or negative Jacobian. Such elements can seriously compromise the quality of results and need to be repaired before an analysis.

Things become even more complicated if the FEM is used for the analysis, as solid elements have to be generated and the mesh has to be truncated at a safe distance. The generation of the mesh is considerably more involved and most mesh generators have to revert to tetrahedral elements. The mesh in Figure 10 would need significant more work to guarantee good quality of the results.

It can be seen that the generation of a mesh, even if the geometry is taken from a CAD program, is a very time consuming part of a simulation and if not done properly will affect the quality of the results.

2.2 Approximation of the unknown

As we have seen in the previous example, the assumption of the approximation of the unknown will also have a considerable influence on the quality of results. In the *isoparametric* approach the description of the geometry and the approximation of the unknown are directly linked, i.e. the same interpolation functions are used for the geometry and the approximation of the unknown. This is not really efficient as in most cases they are not directly linked. For example the geometry of mining excavations is very simple and can be described with few elements but this discretization would not suitable for the approximation of the unknown.

2.3   Solution

The system of equations is usually solved exactly[2] by Gauss elimination (or its variants). For very large systems of equations this is very expensive in terms of computer resources (storage and time) and not really necessary since approximations have occurred in the definition of the geometry and the unknown. Therefore approximate solution techniques have emerged which solve the equations with some error, whose magnitude is in the same order as the errors made in approximating the geometry and the variation of the unknown. These range from iterative methods to H-Matrices, the latter giving not only a considerable reduction in numerical effort but also in storage requirement.

2.4   Recovery of the results

The primary results obtained for solid mechanics problems are displacements at nodal points. The displaced shape can then be obtained by using the approximation functions. In the FEM, stresses are computed from strains that are calculated by taking the derivative of the displacements. Therefore their quality is expected to be less (indeed between elements the satisfaction of equilibrium is no longer guaranteed). Clever schemes were devised in order to get the best possible results by specifying optimal recovery points in a element or other means, generally referred to as *super-convergent stress recovery*. However, there is no hiding the fact that the derived results can only be as good as the approximation used for the primary results (i.e. the displacements).

In the BEM stresses are computed using derived fundamental solutions, which satisfy the conditions of equilibrium and compatibility so the quality of the results is expected to be better, a fact that we will exploit later.

3   A CHANGE OF PARADIGM: TOWARDS A MORE EFFICIENT AND ACCURATE SIMULATION

It is clear from the preceding that currently simulation is far from efficient and safe. The first inefficiency stems from the fact that even if a CAD model is available, the geometry has to be approximated again by a mesh. If an accurate digital model of the geometry is available, why is there a need to approximate it with a mesh?

In a process called *rapid prototyping*, for example, many design modifications need to be made until the optimal design is achieved requiring repeated updates of the mesh. Why do we need to generate a new mesh each time a small design change is made? Would it not be nice if there was a seamless connection with the CAD model and the simulation results could be obtained directly without mesh generation?

The second inefficiency stems from the fact that in the isoparametric approach, the same functions are used to describe the geometry and the approximation of the unknown. This is inefficient, since in many cases there is no direct connection between the two. Consider a problem of very simple shape, which can be described by very few

---

[2]The term is used loosely here and not really correct since round-off errors occur depending on the precision of the storage of numbers in the computer.

elements and linear functions. In most cases this would not be sufficient for describing the variation of the displacements, which subsequently need to be refined, but there is no need to connect this with the definition of the geometry. Furthermore, if higher quality basis functions (such as NURBS introduced later) are used for describing the geometry with very few parameters its description is exact or very accurate, so there is no need to refine the geometry any further but there is a need to refine the approximation of the unknown.

In our journey towards a more efficient simulation we propose a change of paradigm: The geometry description is taken directly from the CAD data and the isoparametric concept is abandoned, i.e. different functions are used for the description of the unknown. To explain how this can be achieved is the main aim of this book.

The CAD community, mainly because they concentrated on geometry, has developed very sophisticated tools for generating a digital representation of it. Whereas the FEM community has been stuck for decades with very simple functions such a Serendipity, Lagrange and in some cases Hermite functions, the CAD community has used for a long time advanced functions such as B-splines and NURBS. The advantage of these functions is that they can describe complex shapes accurately with very few parameters. For example NURBS can describe the geometry of conical surfaces such as cylinders, exactly.

The first task in this book is therefore to make readers familiar with the functions that CAD programs use and how surfaces are described. The second one is to explain the exchange format used to transfer data. Fortunately, a standard has been developed and published at a fairly early stage and most CAD programs adhere to this standard, more or less. As it turns out there is a big advantage of using B-splines or NURBS for the approximation of the unknown too. This is because they exhibit very desirable properties such as good control of continuity and good refinement properties.

4 ORGANIZATION OF THE TEXT

As explained in the preface the book is organized like a road book. The road to knowledge is divided into stages with milestones reached after each stage. Figure 11 shows the roadmap. After the final stage readers should have obtained knowledge of NURBS, how CAD programs use them to define geometry, how an *analysis suitable* description can be obtained and finally see the enormous benefits that can be gained from this technology.

The book is method agnostic, i.e. both FEM and BEM are considered, but the method that is most suitable for the implementation of a seamless integration with CAD will be given preference. Therefore one of the aims is to show the beauty of the Cinderella of numerical methods (the BEM) over its dominant sister the FEM. We start with stage 1 introducing basis functions, i.e. the functions that will be used for describing the geometry and for approximating the unknown. In order to show the subtle differences to currently used functions we start with them and progress to B-splines, NURBS and T-splines. At the end of this stage readers should have a thorough knowledge of NURBS and an idea what T-splines are.

THE JOURNEY TO
ADVANCED SIMULATION

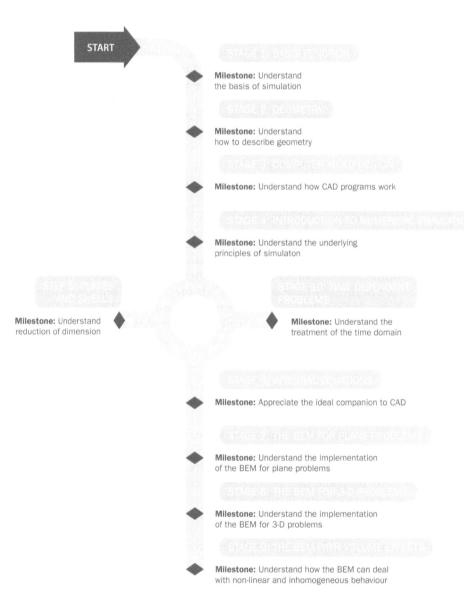

Figure 11 The roadmap to advanced simulation methods.

Stage 2 introduces concepts for defining geometry. Starting with commonly used methods (involving a mesh) readers learn to appreciate the power of NURBS to describe surfaces exactly with very few parameters. Here the concept of trimmed surfaces and their conversion to analysis suitable form is introduced. Infinite surfaces, that are useful for modeling long tunnels and the earth's surface are also introduced here. At the end of this stage readers have the background to understand how geometry can be described with the methods used by CAD programs.

In stage 3 readers become familiar with the concepts of CAD. In particular we look at how such programs describe geometry, at the structure of the data produced by CAD programs and at ways we can extract analysis suitable information. After this stage knowledge of the basic building blocks of simulation should have been obtained.

In stage 4 the concepts of numerical simulation are introduced, starting with a simple one-dimensional analysis, the Bernoulli beam and proceeding to plates with inplane loading. After this stage the underlying principles of simulation and the advantage of using NURBS or B-splines for the approximation of the displacements should be clear.

Stage 5 is a diversion for those readers that are interested in the analysis of Kirchhoff plates and shells. After stage 4 it should be clear that if we proceed to 3-D, a simulation without mesh generation would be only possible if a method is used that relies on a description of surfaces only. Therefore Boundary Integral Equations are introduced in stage 6. These are the predecessors of the BEM and serve well to explain the basic principles. Here we also introduce some novel concepts on how to solve them numerically, such as the Nyström method.

In stage 7 we discuss the implementation of the NURBS based BEM for plane problems and provide test examples in steady state flow and elasticity that clearly show the advantage over domain methods. The 3-D implementation of the BEM is discussed in stage 8, where it is also shown how a simulation can proceed without mesh generation.

If the BEM is only capable of analyzing homogeneous domains with linear material behavior it will have limited application in practice. This topic has unfortunately been neglected for too long by the scientific community. In stage 9 readers will learn how such volume effects can be introduced in the BEM. Some novel and until now untried approaches will be presented. This stage is meant to provide impetus for much needed research on this topic. Finally stage 10 introduces time effects.

Throughout this book mesh generation is not used in the simulation and geometry is either specified by the user or taken directly from CAD data. However, this does not mean that the dream of a seamless integration of CAD and simulation is realized. Far from it, there is still a lot of work to be done. The aim of this book is to encourage more researchers to work in this exciting new area.

BIBLIOGRAPHY

[1] G. Beer, O. Sigl, and J. Brandl. Recent developments and application of the boundary element method. In Petruczczak and Pande, editors, *Numerical models in geomechanics*, pages 461–467. Balkema, 1997.

[2] F.H. Deist, M.D.G. Salamon, and E. Georgiadis. A new digital method for three-dimensional stress analysis in elastic media. *Rock Mechanics*, 5(189–202), 1973.

[3] R.G. Selby, F.J. Vecchio, and M.P. Collins. The failure of an offshore platform. *Concrete International*, 19(8):28–35, 1997.

[4] W. Wappenhaus and J. Richter. Die erste Statik der Welt. *Bautechnik*, 8, 2002.

[5] O.C. Zienkiewicz, D.W. Kelly, and P. Bettess. Marriage a la mode – the best of both worlds (finite elements and boundary integrals). In R. Glowinski, E.Y. Rodin, and O.C. Zienkiewicz, editors, *Energy methods in finite element analysis*, chapter 5, pages 82–107. Wiley, 1979.

Chapter 2

Stage 1: Basis functions

A journey of a thousand miles starts with a single step

Lao-tzu, Chinese philosopher

We start our journey with one of the most important building blocks of numerical simulation, namely functions that can be used to approximate the actual geometry and variation of the unknowns. It can be assumed that every continuous function can be represented as a linear combination of *basis functions*. An introduction to basis functions is presented here. Starting with the first ones, used in the Finite Element method we proceed to the most advanced ones, that we will use later in this book.

1 ONE-DIMENSIONAL BASIS FUNCTIONS

Here we deal with functions that exist in one-dimensional space, either as a function of the coordinate ξ in the case of Lagrange functions or coordinate u in the case of B-splines. The reason for the different naming of the coordinate is to distinguish the different ranges (-1 to 1 and 0 to 1) and to comply as much as possible with published work.

1.1 Lagrange and Serendipity functions

Lagrange and Serendipity functions have been first proposed by Ergatoudis [2] for the FEM and by Lachat [4] for Boundary Element Method. They have been in use since the emergence of numerical modeling software and are still being used today.

Given values of f (f_1 to f_I) at a discrete number of points ξ_1 to ξ_I, we can compute the values of f at any point ξ as

$$f(\xi) = \sum_{i=1}^{I} N_i(\xi) f_i \tag{1}$$

where I is the number of points and the basis functions $N_i(\xi)$ obey the following conditions:

$$N_i(\xi_i) = 1$$
$$N_i(\xi_j) = 0 \quad \text{for } i \neq j \tag{2}$$

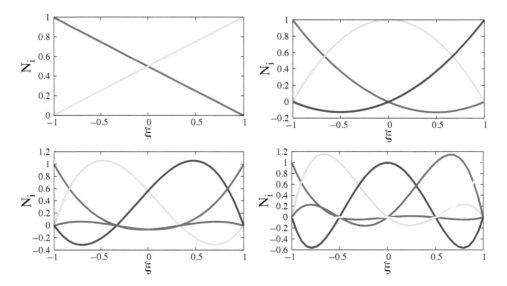

Figure 1 Lagrange functions of order 1 to 4.

is known as the *Kroneker Delta* property and

$$\sum_{i=1}^{I} N_i(\xi) = 1 \tag{3}$$

which is referred to as *partition of unity*.

These conditions are obeyed for example by Lagrange polynomials that are defined as:

$$N_i(\xi) = \prod_{\substack{m \neq i}}^{1 < m \leqslant I} \frac{\xi - \xi_m}{\xi_i - \xi_m} \tag{4}$$

We introduce the *polynomial order* of the function as $p = I - 1$[1].

For a linear interpolation we have $p = 1$, $I = 2$ and

$$N_1(\xi) = \frac{\xi - \xi_2}{\xi_1 - \xi_2}$$

$$N_2(\xi) = \frac{\xi - \xi_1}{\xi_2 - \xi_1} \tag{5}$$

---

[1]The order of a polynomial is the highest exponent of its terms. In some publications this is also referred to as the *degree* of a polynomial. We keep here to the definition in the book *Isogeometric Analysis*.

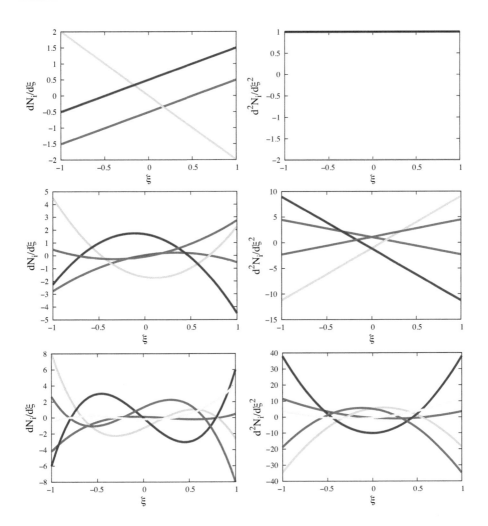

Figure 2 Lagrange function derivatives for orders 2 to 4.

Substitution of the $\xi_1 = -1$ and $\xi_2 = 1$ into equation 5 gives

$$
\begin{aligned}
N_1(\xi) &= \frac{1}{2}(1 - \xi) \\
N_2(\xi) &= \frac{1}{2}(1 + \xi)
\end{aligned}
\tag{6}
$$

For a quadratic interpolation we have $p = 2$ and $I = 3$:

$$
N_1(\xi) = \frac{\xi - \xi_1}{\xi_2 - \xi_1} \cdot \frac{\xi - \xi_3}{\xi_1 - \xi_3}; \quad N_2(\xi) = \frac{\xi - \xi_2}{\xi_1 - \xi_2} \cdot \frac{\xi - \xi_3}{\xi_2 - \xi_3}; \quad N_3(\xi) = \frac{\xi - \xi_1}{\xi_3 - \xi_1} \cdot \frac{\xi - \xi_2}{\xi_3 - \xi_1}
\tag{7}
$$

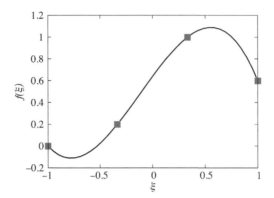

Figure 3   Example of interpolation between 4 points with Lagrange polynomials of order 3.

Substitution of $\xi_1 = -1$, $\xi_2 = 1$ and $\xi_3 = 0$ gives

$$N_1(\xi) = \frac{\xi - 1}{2} \cdot \xi; \quad N_2(\xi) = \frac{\xi + 1}{2} \cdot \xi; \quad N_3(\xi) = (\xi + 1) \cdot (\xi - 1) \tag{8}$$

The first derivative to ξ is given by:

$$\frac{\partial N_i(\xi)}{\partial \xi} = \sum_{j=1:I}^{j \neq i} \left(\frac{1}{\xi_i - \xi_j} \prod_{m=1:I}^{m \neq j, m \neq i} \frac{\xi - \xi_m}{\xi_i - \xi_m} \right) \tag{9}$$

At the time of the emergence of the FEM it was proposed, that functions which obey conditions (2) and (3) can be found in a non-mathematical way, i.e. without using the product formula (4).

The functions were subsequently called *Serendipity* functions[2]. For one-dimensional basis functions there is no difference between Serendipity and Lagrange functions, except for the way they are determined and how the nodes are numbered (i.e. for Serendipity functions the edge nodes are numbered first, for reasons to be revealed later). For linear Serendipity functions the derivatives are given by:

$$\frac{\partial N_1(\xi)}{\partial \xi} = -0.5; \quad \frac{\partial N_2(\xi)}{\partial \xi} = 0.5 \tag{10}$$

and for quadratic functions:

$$\frac{\partial N_1(\xi)}{\partial \xi} = \xi - 0.5; \quad \frac{\partial N_2(\xi)}{\partial \xi} = \xi + 0.5; \quad \frac{\partial N_3(\xi)}{\partial \xi} = -2 \cdot \xi \tag{11}$$

---

[2]Serendipity means a "pleasant surprise". It was first coined by Horace Walpole in 1754. In a letter he wrote to a friend Walpole explained an unexpected discovery he had made by reference to a Persian fairy tale, The Three Princes of Serendip. The princes, he told his correspondent, were *always making discoveries, by accidents and sagacity, of things which they were not in quest of.*

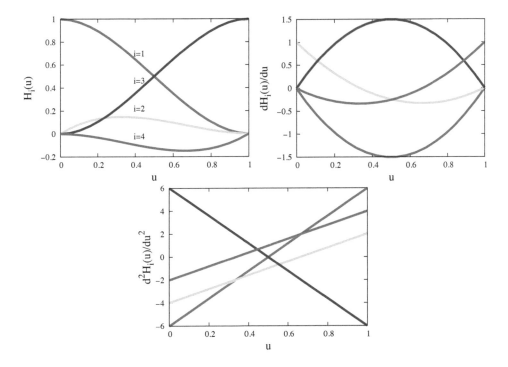

Figure 4 Hermite polynomials and their derivatives.

The second derivative is given by

$$
\frac{\partial^2 N_i(\xi)}{\partial \xi^2} = \sum_{\substack{j=1:I}}^{j\neq i} \frac{1}{\xi_i - \xi_j} \cdot \left(\sum_{\substack{n=1:I}}^{n\neq j, n\neq i} \left(\frac{1}{\xi_i - \xi_n} \cdot \prod_{\substack{m=1:I}}^{m\neq j, m\neq i, m\neq n} \frac{\xi - \xi_m}{\xi_i - \xi_m} \right) \right) \tag{12}
$$

For a quadratic Serendipity function we have

$$
\frac{\partial^2 N_1}{\partial \xi^2} = 1; \quad \frac{\partial^2 N_2}{\partial \xi^2} = 1; \quad \frac{\partial^2 N_3}{\partial \xi^2} = -2 \tag{13}
$$

Figure 1 shows Lagrange functions of order 1 to 4. It can be observed that the higher order functions have significant negative values and exhibit oscillatory behavior. Figure 2 shows the derivatives for functions of order 2 and 3.

Figure 3 shows the interpolation between given values of function $f = (0, 0.2, 1.0, 0.6)$ at points $\xi = (-1, -0.3, 0.3, 1)$ with Lagrange polynomials of order $p = 4$.

Two things can be observed:

- The curve goes through the given values of the function (nodal points).
- There is little control on what the curve does between the nodal points.

It turns out that these functions are not very well suited for numerical simulation. The main problem is that one has little control of what happens between the nodal points. In fact the oscillatory tendency of higher order functions has been found to be a problem in simulation, especially when the function to be approximated varies rapidly. One step towards achieving more control over the shape is Hermitian cubic polynomials.

Hermite polynomials The functions are defined in a local coordinate u that goes from 0 to 1 by:

$$H_1(u) = 2u^3 - 3u^2 + 1; \quad H_2(u) = L \cdot (u^3 - 2u^2 + u)$$
$$H_3(u) = -2u^3 + 3u^2; \quad H_4(u) = L \cdot (u^3 - u^2) \tag{14}$$

where L is the length of the interpolation region.

The first derivative is

$$\frac{dH_1(u)}{du} = 6u^2 - 6u; \quad \frac{dH_2(u)}{du} = L \cdot (3u^2 - 4u + 1)$$
$$\frac{dH_3(u)}{du} = -6u^2 + 6u; \quad \frac{dH_4(u)}{du} = L \cdot (3u^2 - 2u) \tag{15}$$

The second derivative is

$$\frac{d^2H_1(u)}{du^2} = 12u - 6; \quad \frac{d^2H_2(u)}{du^2} = L \cdot (6u - 4)$$
$$\frac{d^2H_3(u)}{du^2} = -12u + 6; \quad \frac{d^2H_4(u)}{du^2} = L \cdot (6u - 2) \tag{16}$$

The functions and derivatives are plotted in Figure 4.

A function may now be approximated by

$$f(u) = H_1(u) \cdot f_1 + H_1(u) \cdot f_1' + H_3(u) \cdot f_2 + H_1(u) \cdot f_2' \tag{17}$$

where f_1, f_1' and f_2, f_2' are function values and derivatives at points 1 and point 2.

1.2 From B-splines to NURBS

Bernstein polynomials The predecessors of B-splines are Bernstein[3] polynomials (see [1] and [6]) which are defined by

$$B_{i,p} = c_{i,p} \cdot u^i \cdot (1 - u)^{p-i-1} \tag{18}$$

where u is the local coordinate, $c_{i,p}$ is a binomial coefficient and p is the order of the function.

---

[3]Sergei Natanovich Bernstein (1880 to 1968) was a Russian mathematician known for contributions to partial differential equations, differential geometry, probability theory, and approximation theory.

Figure 5 Sergei Natanovich Bernstein.

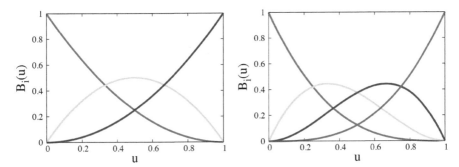

Figure 6 Bernstein polynomials of order 2 and 3.

The Bernstein polynomials for orders $p = 1$ to $p = 3$ are

$$B_{0,1} = (1 - u), \quad B_{1,1} = u \tag{19}$$

$$B_{0,2} = (1 - u)^2, \quad B_{1,2} = 2u(1 - u), \quad B_{2,2} = u^2 \tag{20}$$

$$B_{0,3} = (1 - u)^3, \quad B_{1,3} = 2u(1 - u)^2, \quad B_{2,3} = 3u^2(1 - u), \quad B_{3,3} = 3u^3 \tag{21}$$

The function values can be approximated by[4]

$$f(u) = \sum_{i=0}^{I} B_{i,p}(u) \cdot P_i \tag{22}$$

where P_i are parameter values at control points. This is major difference to the functions presented previously. It means that we define the shape of the curve without interpolating between points. This also means that these points may no longer be on the curve and therefore are not *interpolatory*.

---

[4]To comply with published work we sum from zero to $I = p + 1$ but this will be changed later for programming purposes, since a zero index is not allowed in Octave/MATLAB.

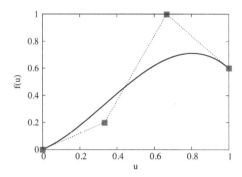

Figure 7 Example of approximation of curve with 4 control points and with Bernstein polynomials of order 3.

The Bernstein polynomials of order 1 are identical to the Lagrange functions. Figure 6 shows the Bernstein polynomials of order 2 and 3. It can be noted that the polynomials no longer have the *Kroneker-Delta* property, i.e. the functions do not in general have unit and zero values at points. We also note that the functions only have positive values and this prevents any oscillatory behavior, experienced previously.

We show the approximation of a curve using the same points as in Figure 3 but this time we treat the points where the function value is specified as **control points**. We see that there is quite a difference. The curve only goes through the specified points at the beginning and the end.

The polygon connecting the control points is also called a *control polygon*. Although this gives better control over the shape of the curve, it would be also desirable to be able to control the continuity of the basis functions. This can be achieved by B-splines.

B-splines The further development of Bernstein polynomials can be attributed to Bezier[5] and de Boor. To define B-splines we start with a *knot vector*. This is a vector containing a series of non-decreasing values of the local coordinate:

$$\Xi = (u_0 \quad u_1 \quad \cdots \quad u_N) \tag{23}$$

We define the entries in the vector as *knots* and the interval between knots as the *knot span*. With these entries in the knot vector a recursive formula is applied for computing the functions.

We explain the generation of B-splines in Figure 9 for a knot vector

$$\Xi = (u_0 = 1, \ u_1 = 2, \ u_3 = 3, \ \ldots).$$

---

[5] Pierre Étienne Bézier (1910 to 1999) was a French engineer and one of the founders of the fields of solid, geometric and physical modeling as well as in the field of representing curves, especially in CAD/CAM systems. As an engineer at Renault, he became a leader in the transformation of design and manufacturing, through mathematics and computing tools, into computer-aided design and three-dimensional modeling.

Figure 8   Pierre Étienne Bézier.

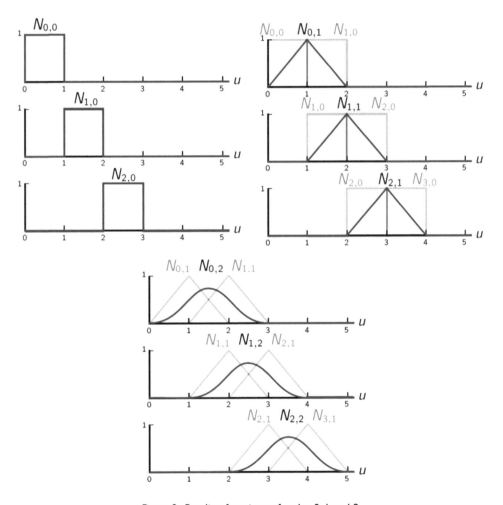

Figure 9   B-spline functions of order 0, 1 and 2.

First we compute the functions for order $p = 0$ (constant) and for $i = 0, \ldots, N$.

$$N_{i,0}(u) = \begin{cases} 1 & \text{if } u_i \leqslant u < u_{i+1} \\ 0 & \text{otherwise} \end{cases} \tag{24}$$

Higher order basis functions are defined by referencing lower order functions:

$$N_{i,p}(u) = \frac{u - u_i}{u_{i+p} - u_i} N_{i,p-1}(u) + \frac{u_{i+p+1} - u}{u_{i+p+1} - u_{i+1}} N_{i+1,p-1}(u) \tag{25}$$

It can be seen that the higher order functions are a linear combination of the lower order functions. Note that some functions are zero along some portions of the local coordinate space, i.e. have limited span.

The coordinates in the *knot vector* may be non-uniformly spaced and we refer to the corresponding functions as *non-uniform B-splines*.

The first derivatives are computed by

$$N'_{i,p}(u) = \frac{d}{du} N_{i,p}(u) = \frac{p}{u_{i+p} - u_i} N_{i,p-1}(u) - \frac{p}{u_{i+p+1} - u_{i+1}} N_{i+1,p-1}(u) \tag{26}$$

The second derivatives are computed by:

$$\frac{d^2}{du^2} N_{i,p}(u) = \frac{p}{u_{i+p} - u} \cdot \frac{d}{du} N_{i,p-1}(u) - \frac{p}{u_{i+p+1} - u_{i+1}} \cdot \frac{d}{du} N_{i+1,p-1}(u) \tag{27}$$

The knot vector can also be used to control the continuity of the functions. Function continuity can be controlled by a repetition of values in the knot vector. A repetition of the values $p + 1$ times means C^{-1} continuity (i.e. dis-continuity) and a repetition p times means C^0 continuity (i.e. continuity of function values only, not tangents).

In the following we will consider only *open knot vectors* where the values are repeated $p + 1$ times at the beginning and the end[6]. Furthermore, to simplify maters we will work with a parameter space that spans from 0 to 1, so that the entries in the knot vector u_i will be $\geqslant 0$ and $\leqslant 1$. Indeed the functions of the NURBS toolkit only work with this range and knot vectors provided as input, that not comply to this will be normalised.

Let us construct the functions for the knot vector $\Xi = (u_0 = 0, u_1 = 0, u_2 = 0, u_3 = 1, u_4 = 1, u_5 = 1)$. In Figure 10 we show the construction of the basis functions of order 0 and 1 in the index space, i.e. the numbers along the axis denote the indexes of the entries in the knot vector, and not the local coordinate (u) as was the case in the previous example, where both were the same. The parameter space is a subspace of the index space that ranges from $u_2 = 0$ to $u_3 = 1$.

First we compute the functions for $p = 0$. We can easily see from Equation (24) that only the function $N_{2,0} = 1$ exists in the parameter space.

---

[6]The NURBS toolkit, which we will use in this book, only works with open knot vectors. However as we will see later this is not a restriction.

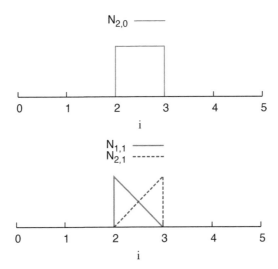

Figure 10 Construction of B-spline functions of orders 0 and 1 in the index space for knot vector $\Xi = (0, 0, 0, 1, 1, 1)$.

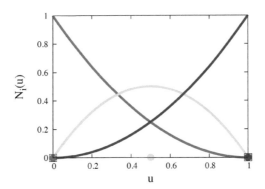

Figure 11 Resulting B-spline functions (knots are depicted by squares and anchors by filled circles).

The following non-zero basis functions for $p = 1$ can be computed using Equation (25)

$$N_{1,1} = (1 - u) \cdot N_{2,0}; \quad N_{2,1} = u \cdot N_{2,0} \tag{28}$$

Finally the basis functions for $p = 2$ can be obtained

$$N_{0,2} = (1 - u) \cdot N_{1,1}; \quad N_{1,2} = u \cdot N_{1,1} + (1 - u) \cdot N_{2,1};$$
$$N_{2,2} = u \cdot N_{2,1} \tag{29}$$

The computed functions are continuous between $u = 0$ and $u = 1$ (see Figure 11).

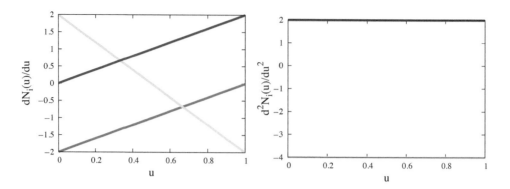

Figure 12 B-spline functions and derivatives for $p = 2$.

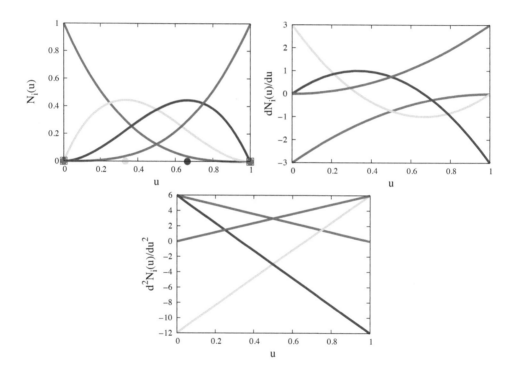

Figure 13 B-spline functions and derivatives of order 3 for knot vector $\Xi = (0, 0, 0, 0, 1, 1, 1, 1)$ (knots are depicted by squares and anchors by filled circles).

We note that for the case of Lagrange polynomials each basis function was directly associated with a nodal point, due to its Kronecker delta property. Since B-splines do not have this property *anchors* are introduced that connect each basis function to a position in the parameter space.

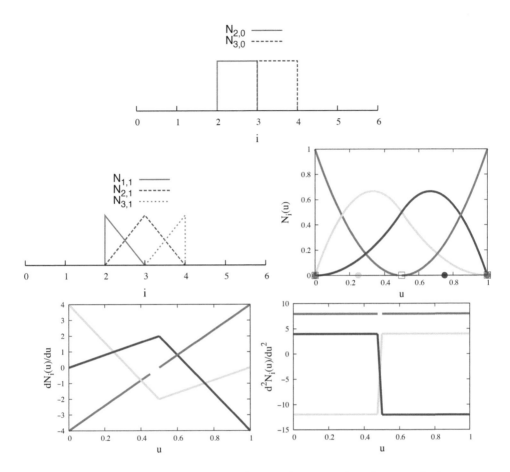

Figure 14 Construction of B-spline functions of orders 0 and 1 in the index space for knot vector $\Xi = (0, 0, 0, 0.5, 1, 1, 1)$. Resulting basis function and derivatives (knots are depicted by squares and anchors are depicted by filled circles).

The location of the i-th anchor (a_i) in parameter space can be computed using a formula by Greville [3]:

$$u(a_i) = \frac{u_{i+1} + u_{i+2} + \cdots + u_{i+p}}{p} \qquad i = 0, 1, \ldots, I \tag{30}$$

The locations of the anchors are shown as color coded filled circles, together with the basis functions. The derivatives of the functions are shown in Figure 12. B-splines and derivatives for order 3 are shown in Figure 13.

We can then make the following modifications:

1 Insertion of one knot at $u = 0.5$: We can easily see from Equation (24) that only functions $N_{2,0} = 1$ and $N_{3,0} = 1$ exist in the parameter space. The following

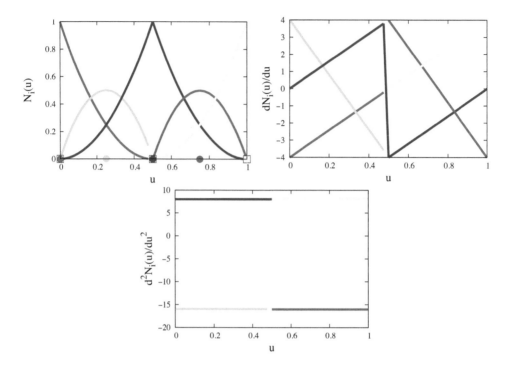

Figure 15 B-spline functions and derivatives for Knot vector $\Xi = (0, 0, 0, 0.5, 0.5, 1, 1, 1)$ (knots are depicted by squares and anchors are depicted by filled circles).

non-zero basis functions for $p = 1$ can be computed using Equation (25)

$$N_{1,1} = (1 - u) \cdot N_{2,0} \tag{31}$$
$$N_{2,1} = u \cdot N_{2,0} + 2 \cdot (1 - u) \cdot N_{3,0}$$
$$N_{3,1} = (2u - 1) \cdot N_{3,0}$$

These intermediate constructions and the final result are shown in Figure 14. We can see that functions $N_{0,1}$ and $N_{4,1}$ only span half of the parameter space.

2 Insertion of two knots at $u = 0.5$: This means that the blue basis function has only C^0 continuity (see Figure 15) and all other basis functions only span half of the parameter space. For this case, two knots are located at the same location. It is noted that the same result can be achieved if the parameter space is split into two and Bernstein polynomials are used for each subspace.

We re-plot the function in Figure 7 with the modified knot vectors in Figures 16 and 17. It can be seen that the shape of the curve has been altered when compared to Bernstein polynomials. In the second case the curve has only a C^0 continuity at $u = 0.5$.

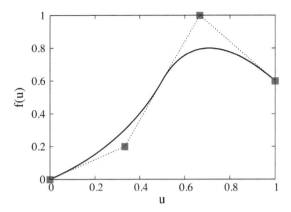

Figure 16 Plot with the same control points as in Figure 7 but with knot vector $\Xi = (0, 0, 0, 0.5, 1, 1, 1)$.

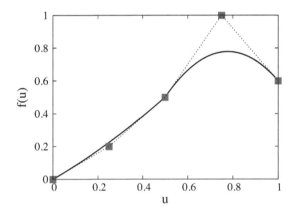

Figure 17 Plot with the same control points as in Figure 7 plus an additional one and with knot vector $\Xi = (0, 0, 0, 0.5, 0.5, 1, 1, 1)$.

NURBS The fact that we can control continuity and span of the functions makes them very powerful. However, further improvements are possible. The shape of the curve can be further controlled by using weights.

The basis function becomes:

$$R_{i,p}(u) = \frac{N_{i,p}(u)w_i}{\sum_{j=0}^{I} N_{j,p}(u)w_j} \tag{32}$$

where $N_{i,p}(u)$ are the B-splines defined previously and w_i are weights.

To retain the *partition of unity* condition a division by the sum of the B-splines times the weights is required.

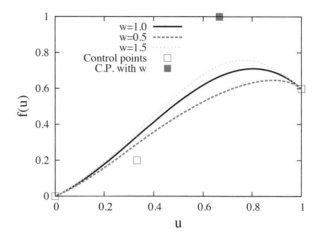

Figure 18 A function approximated with NURBS basis functions and different values of weight for third control point.

Function $R_{i,p}(u)$ is no longer a polynomial, but a *rational* function. Therefore the function just defined is also called non-uniform rational B-spline or NURBS.

The first derivative of the NURBS functions are given by:

$$\frac{d}{du}R_{i,p}(u) = w_i \frac{W(u)N'_{i,p}(u) - W'(u)N_{i,p}(u)}{W^2(u)} \tag{33}$$

with

$$W(u) = \sum_{j=0}^{I} N_{j,p}(u)w_j \tag{34}$$

$$W'(u) = \sum_{j=1}^{I} N'_{j,p}(u)w_j \tag{35}$$

A function can now be approximated by

$$f(u) = \sum_{i=0}^{I} R_{i,p}(u) \cdot P_i \tag{36}$$

To see the effect of the weights, we plot a curve with different values of w associated with the third control point in Figure 18. It can be seen that a higher value of weight draws the curve towards the control point. It is this additional flexibility that makes the NURBS so attractive for the description of the geometry, as will be seen later.

2 TWO-DIMENSIONAL BASIS FUNCTIONS

Here we discuss basis functions that exist in two-dimensional space. To comply with published work, we choose the coordinates ξ, η for the Lagrange polynomials and u, v for NURBS.

2.1 Lagrange and Serendipity functions

Lagrange polynomials   For the Lagrange functions a tensor product of functions in ξ and η directions is used:

$$N_{ij}(\xi, \eta) = N_i(\xi) \cdot N_j(\eta) \tag{37}$$

where $N_i(\xi)$ has been defined previously and

$$N_j(\eta) = \prod_{\substack{m \neq j}}^{1 < m \leqslant J} \frac{\eta - \eta_m}{\eta_j - \eta_m} \tag{38}$$

The derivatives are

$$\frac{\partial N_{ij}(\xi, \eta)}{\partial \xi} = \frac{\partial N_i(\xi)}{\partial \xi} \cdot N_j(\eta) \tag{39}$$

$$\frac{\partial N_{ij}(\xi, \eta)}{\partial \eta} = N_i(\xi) \cdot \frac{\partial N_j(\eta)}{\partial \eta} \tag{40}$$

The Lagrange shape functions are depicted in Figures 19 and 20.

Serendipity functions   For one-dimensional basis functions there was no difference to Lagrange functions. However, this is not the case for the two-dimensional case.

Serendipty functions for $p = 2$, are not derived using a tensor product, only have 8 values (function $N_{2,2}$, the *bubble function* is missing) and are given by:

$$N_5 = \frac{1}{2}(1 - \xi^2)(1 - \eta) \tag{41}$$

$$N_6 = \frac{1}{2}(1 - \eta^2)(1 - \xi) \tag{42}$$

$$N_7 = \frac{1}{2}(1 - \xi^2)(1 + \eta) \tag{43}$$

$$N_8 = \frac{1}{2}(1 - \eta^2)(1 + \xi) \tag{44}$$

$$N_1 = \frac{1}{4}(1 - \xi)(1 - \eta) - \frac{1}{2}N_5 - \frac{1}{2}N_8 \tag{45}$$

$$N_2 = \frac{1}{4}(1 + \xi)(1 - \eta) - \frac{1}{2}N_5 - \frac{1}{2}N_6 \tag{46}$$

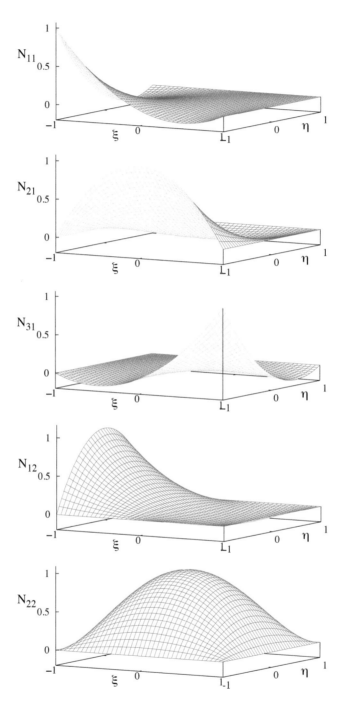

Figure 19 Lagrange shape functions.

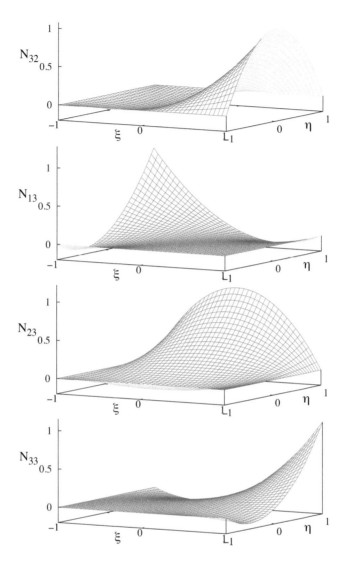

Figure 20 Figure 19 continued.

$$N_3 = \frac{1}{4}(1 + \xi)(1 + \eta) - \frac{1}{2}N_6 - \frac{1}{2}N_7 \tag{47}$$

$$N_4 = \frac{1}{4}(1 - \xi)(1 + \eta) - \frac{1}{2}N_7 - \frac{1}{2}N_8 \tag{48}$$

One advantage of Serendipity functions is that each one of the *midside node* functions (5 to 8) can be set to zero, resulting in a mixed linear-parabolic interpolation. It can be seen in Figure 21 that the functions do not have zero values at $\xi = 0$, $\eta = 0$ as the Lagrange functions do.

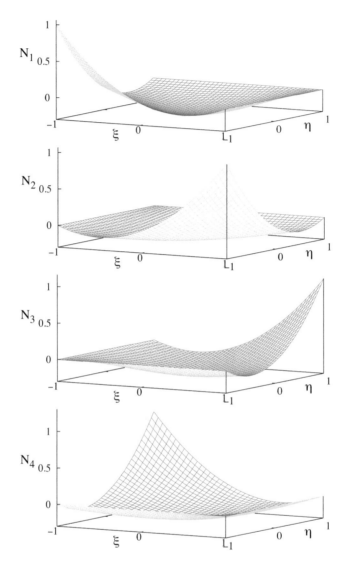

Figure 21 Serendipity shape functions.

The derivatives of the Serendipity functions to ξ are given by

$$\frac{\partial N_5}{\partial \xi} = \xi \cdot (1 - \eta) \tag{49}$$

$$\frac{\partial N_6}{\partial \xi} = \frac{1}{2} \cdot \xi \cdot (1 - \eta^2) \tag{50}$$

$$\frac{\partial N_7}{\partial \xi} = -\xi \cdot (1 + \eta) \tag{51}$$

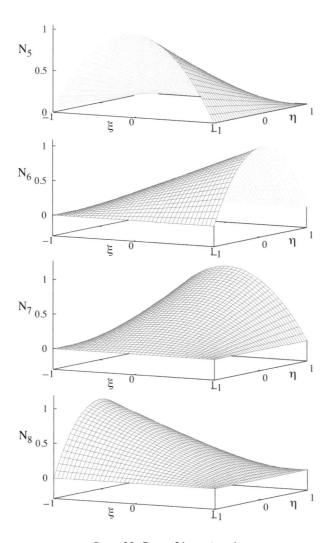

Figure 22 Figure 21 continued.

$$\frac{\partial N_8}{\partial \xi} = \frac{1}{2} \cdot \xi \cdot (1 - \eta^2) \tag{52}$$

$$\frac{\partial N_1}{\partial \xi} = -\frac{1}{4} \cdot (1 - \eta) - \frac{1}{2} \frac{\partial N_5}{\partial \xi} - \frac{1}{2} \frac{\partial N_8}{\partial \xi} \tag{53}$$

$$\frac{\partial N_2}{\partial \xi} = \frac{1}{4} \cdot (1 - \eta) - \frac{1}{2} \frac{\partial N_5}{\partial \xi} - \frac{1}{2} \frac{\partial N_6}{\partial \xi} \tag{54}$$

$$\frac{\partial N_3}{\partial \xi} = -\frac{1}{4} \cdot (1 + \eta) - \frac{1}{2} \frac{\partial N_6}{\partial \xi} - \frac{1}{2} \frac{\partial N_7}{\partial \xi} \tag{55}$$

$$\frac{\partial N_4}{\partial \xi} = -\frac{1}{4} \cdot (1 + \eta) - \frac{1}{2} \frac{\partial N_7}{\partial \xi} - \frac{1}{2} \frac{\partial N_8}{\partial \xi} \tag{56}$$

2.2 B-splines

Two-dimensional B-splines are also computed using a tensor product:

$$N_{i,j}^{p,q} = N_{i,p}(u) \cdot N_{j,q}(v) \tag{57}$$

where $N_{i,p}(u)$ has been introduced previously and $N_{j,q}(v)$ is a B-spline of order q in the v direction. We now have 2 knot vectors: one in u- and the other in v-direction.

Consider B-splines of order $p = q = 2$ and the Open Knot vectors:

$$\Xi_u = (0, 0, 0, 1, 1, 1) \tag{58}$$

$$\Xi_v = (0, 0, 0, 1, 1, 1) \tag{59}$$

There are 9 B-spline functions which are plotted in Figures 23 and 24. The main difference to the Lagrange functions is that all values are positive. As with the one-dimensional functions we can increase the order of the basis function. Figure 25 shows some basis functions for the case where the order in u-direction (p) is increased to 3.

We can also perform knot insertion. Figure 26 shows the basis functions for $p = q = 1$ and a knot inserted at $(u = 0.5, v = 0.5)$. We see that some functions have a *local support*, i.e. are nonzero only over a portion of the parameter space.

2.3 NURBS

NURBS functions are generated by multiplying the B-splines with weights and ensuring the partition of unity condition:

$$R_{i,j}^{p,q} = \frac{N_{i,p}(u) \cdot N_{j,q}(v) \cdot w_{i,j}}{W} \tag{60}$$

with

$$W = \sum_{j=0}^{J} \sum_{i=0}^{I} N_{i,p}(u) \cdot N_{j,q}(v) \cdot w_{i,j} \tag{61}$$

$N_{i,p}(u)$ and $N_{j,q}(v)$ are B-spline functions of local coordinates u or v of order p or q and $w_{i,j}$ are weights. The only difference to B-splines is that the amplitude of the basis functions is determined by the weights. The derivative to u for example is given by:

$$\frac{d}{du} R_{i,j}^{p,q}(u, v) = \frac{N' w_{i,j} - W'(u) \cdot R_{i,j}^{p,q}(u, v)}{W^2(u)} \tag{62}$$

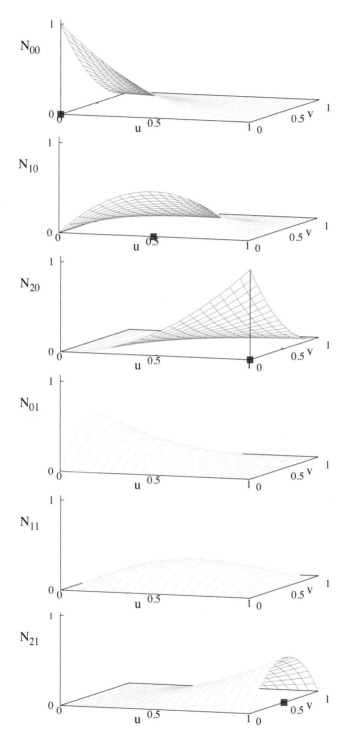

Figure 23 B-spline functions of order $p = q = 2$ and with Knot vectors $\Xi_u = (0, 0, 0, 1, 1, 1)$ and $\Xi_v = (0, 0, 0, 1, 1, 1)$ with associated anchor locations in the u, v coordinate system (some anchors are hidden from view).

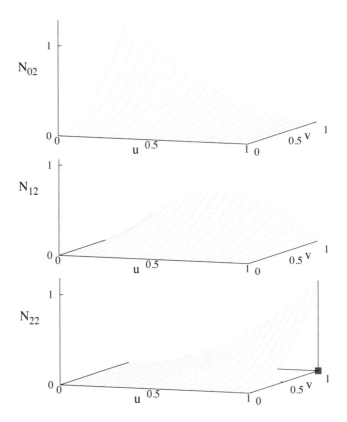

Figure 24 Figure 23 continued.

with

$$W(u) = \sum_{\hat{i}=0}^{I} \sum_{\hat{j}=0}^{J} N_{\hat{i},p}(u) \cdot N_{\hat{j},q}(v) w_{\hat{i},\hat{j}} \tag{63}$$

$$W'(u) = \sum_{\hat{i}=0}^{I} \sum_{\hat{j}=0}^{J} N'_{\hat{i},p}(u) \cdot N_{\hat{j},q}(v) w_{\hat{i},\hat{j}} \tag{64}$$

and

$$N' = \frac{d}{du}(N_{i,p}(u,v)) \cdot N_{j,q}(u,v) \tag{65}$$

2.4 T-splines

One problem with tensor product of NURBS is that the number of control points has to be equal at opposing sides. In some cases it would be convenient to have different

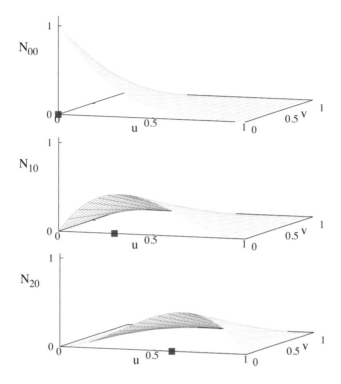

Figure 25 Some B-spline functions of order $p=3, q=1$ and with knot vectors $\Xi_u = (0,0,0,0,1,1,1,1)$ and $\Xi_v = (0,0,1,1)$ with associated anchor locations in the u, v coordinate system.

number of control points on opposing sides, for example if a local refinement of the geometry description is necessary. The solution to this problem is T-splines (see [8] and [7]). Only a simple explanation of T-splines is given here. Interested readers are referred to the quoted literature. To simply explain T-splines consider the basis functions in Figure 26. They have 3 control points on each side. Let us assume that we want to reduce the number of control points on the side $u = 0$ to two as shown in Figure 27. For the border $u = 0$ we have a changed knot vector ($\Xi_v = (0,0,1,1)$).

To get away from the restrictions of the tensor product we introduce the concept of *local anchors*. Local anchors are locations in the parametric space associated with individual basis functions of a T-spline. If the order of the function is odd the location of the anchors are at the knots. For an even order the locations are at the center of knot spans. In the example presented here the order is odd, the location of the anchors is at the knots and is shown in Figure 27.

For each anchor point we extract *local knot vectors* in the u and v direction (in Figure 27 these are the values surrounded by a red rectangle) and compute local shape functions $N_i^u(u)$ and $N_i^v(v)$ that are now based on the extracted local knot vectors. As an example we show in Figure 28 the numbering of anchors and the local basis functions $N_2^u(u)$ and $N_2^v(v)$.

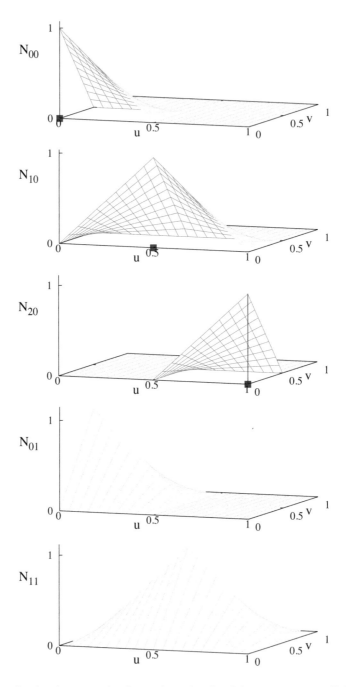

Figure 26 Some B-spline functions of order $p = 1, q = 1$ and with knot vectors $\Xi_u = (0, 0, 0.5, 1, 1)$ and
$\Xi_v = (0, 0, 0.5, 1, 1)$ with associated anchor locations in the u, v coordinate system.

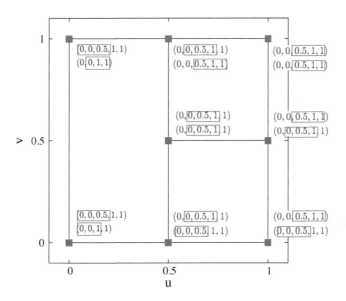

Figure 27 Location of anchor points with extracted local knot vectors.

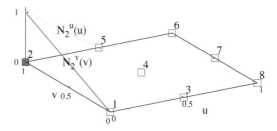

Figure 28 Numbering of anchors and local basis functions for anchor 2.

The basis functions are then given by

$$R_i(u,v) = \frac{N_i^{uv}(u,v) \cdot w_i}{\sum_{j=1}^{m} N_j^{uv}(u,v) \cdot w_j} \tag{66}$$

with $N_i^{uv}(u,v) = N_i^{u}(u) \cdot N_i^{v}(v)$.

Some of the resulting T-spline basis functions are shown in Figure 29. It can be seen that the basis functions associated with anchors on $u = 0$ now span from $v = 0$ to 1 instead of from 0 to 0.5 as was the case when using a tensor product.

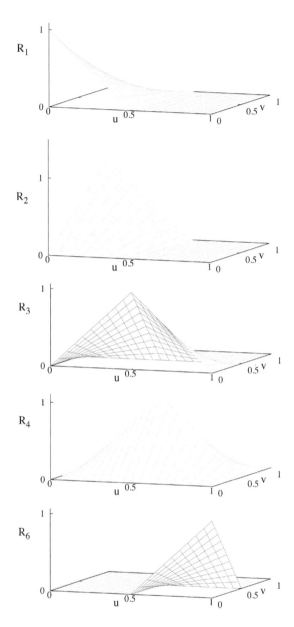

Figure 29 Some T-spline basis functions.

3 PROGRAMMING

Since the emphasis of the book is on understanding concepts and implementation a few words about programming are in order. MATLAB or its freeware alternative OCTAVE is a higher level programming language that is ideally suited for quick implementation and trying out of concepts. Indeed, the author believes that a better understanding of concepts can be gained by a graphic representation as most engineers understand mathematical concepts better using computer graphics. Higher level programming languages are ideally suited for this task as they allow quick and uncomplicated programming as well as the creation of graphical output. Throughout the book software has been developed that allows a better understanding of the concepts, the degree of accuracy that can be obtained and the convergence characteristics of the methods. OCTAVE was mainly chosen because it is available free of charge and because the NURBS toolkit was available in this language shortening the time and effort. For the purposes of this book the more sophisticated aspects of MATLAB were not required. However, all programs developed should run in MATLAB with no or only minor modifications.

In the following we first introduce the NURBS toolkit. Since the documentation is rather sparse, a whole section is devoted to the explanation of the input and output. Although, one has to get used to some aspects of the toolkit, it is efficiently programmed and bug free. As a reminder, we mention that the convention is that the first subscript in an array denotes row number and the second column number.

4 THE NURBS TOOLKIT

The toolkit [9] can be downloaded free. Here we only introduce the OCTAVE functions for determining the values of NURBS at parametric points. Other functions will be introduced in the next chapters as required. A few remarks on the software are appropriate here. The functions take *homogeneous coordinates/parameters* at the control points as input. **This means that they have to be multiplied by the weights before being supplied to the function.** The functions assume that the entries in knot vector range from 0 to 1. Knot vectors can be supplied with different range of entries but are changed internally.

The first function introduced is *nrbmak*, which makes a scalar structure containing all the information required to define a NURBS: The knot vector and the control point coordinates/parameters and weights for a NURBS curve or surface. The number of control points and the order of the basis functions is deduced from the size of the array supplied in the input. Care has to be taken when these arrays are reused with different sizes (in this case a *clear* command must be used). The toolkit always demands 3 coordinates. For plane curves the z-coordinate is set to zero. As will be pointed out in the detailed description of the toolkit, there are a number of inconsistencies. For example different definitions of order or degree of the basis functions are used. Also the numbering of the basis functions starts from zero in one place and from one in others.

```
function nrbmak(coefs, knots)
%---------------------------------------------------
% Makes a structure of the information required to define a NURBS.
%  Input:
%  coefs  ...  weighted control point coordinates and weights
%  For NURBS curves:
%  Array of dimension (4, I), weighted coordinates, weights of control points
%  coefs(1:4, i)= xi*wi, yi*wi,  zi*wi, wi
```

```
%  For NURBS surfaces:
%  Array of dimension (4, I, J)
%  coefs(1:4, i, j)= xij*wij, yij,*wij, zij*wij, wij
%
%  knots   ...  knot vector(s)
%  For NURBS curves:
%  knotu  ... knot vector in u-direction
%  For NURBS surfaces:
%  {knotu, knotv}  ... knot vectors in u,v direction
%
%  Output:
%  nurbs ... NURBS structure
%  nurbs.number  ...  Number of control points
%  nurbs.coefs   ...  Coefficients
%  nurbs.order [u-order, v-order] ... orders of NURBS in u and v directions +1
%  nurbs.knots {knotu, knotv}   ...   knot vectors in u and v directions
%--------------------------------------------------
```

Example NURBS curve:

```
nurbs= nrbmak([0 0.707 0;0 0.707 1;0 0 0; 1 0.707 1],[0 0 0 1 1 1])
nurbs = scalar structure containing the fields:
    form = B-NURBS
    dim =  4
    number =  3
    coefs =
        0.00000    0.70700    0.00000
        0.00000    0.70700    1.00000
        0.00000    0.00000    0.00000
        1.00000    0.70700    1.00000
    order =  3
    knots =
        0    0    0    1    1    1
```

Example NURBS surface:

```
coefs(:,:,1)= [0 0;1 0;0 0;1 1]
coefs(:,:,2)= [0 1;1 1;0 0;1 1]
nurbs=nrbmak(coefs,{[0 0 1 1],[0 0 1 1]})
nurbs =
  scalar structure containing the fields:
    form = B-NURBS
    dim =  4
    number =
        2    2
    coefs =
    ans(:,:,1) =
```

```
     0   0
     1   0
     0   0
     1   1
ans(:,:,2) =
     0   1
     1   1
     0   0
     1   1
knots =
{
  [1,1] =
       0   0   1   1
  [1,2] =
       0   0   1   1
}
order =
     2   2
```

Notes: For the coefficients always 4 values must be specified. The first 3 are the coordinates or parameters of the control point multiplied by the weight and the last is the weight. Control point parameters must be always supplied, even if only the value of the basis function is required. In this case a zero value for all parameters may be given, with the weights specified. For some reason two definitions are used in the toolkit for the order of the basis function. In function nrbmak the meaning of *order* is actually one higher than defined in this chapter. In function findspan the meaning of *degree* is identical to the meaning of *order* defined in this book.

The following function determines the span of a basis function. This will be required later for determining which basis functions have a non-zero value at a parametric point.

```
function s = findspan(n,p,u,knot)
%-----------------------------------------
% Find the span of a B-Spline at a parametric point
%
%  Input:
%    n - number of control points - 1
%    p - spline degree = order
%    u - parametric point
%    knot - knot vector
%
%  Output:
%    s - knot span index
%-----------------------------------------
```

Example:

```
s= findspan(2,2,0.5,[0 0 0 1 1 1])
s =  2
```

The following function determines the numbers of the basis functions that are not zero at a parametric point. The basis functions are numbered beginning with zero here.

```
function idx = numbasisfun (s, u, p, knot)
%-------------------------------------------------
% List numbers of non-zero basis functions in a given knot-span
%
%    Input:
%       s - knot span  ( from findspan)
%       u - parametric point
%       p - spline degree
%       knot - knot vector
%
%    Output:
%       idx ... index to non-zero basis functions
% -------------------------------------------------
```

Example

```
idx= numbasisfun(s,0.5,2,[0 0 0 1 1 1])
idx = 0   1   2
```

The following function will compute the basis function value at parametric points (u, v). the values of the points can be supplied as individual points or as points on a grid. In the latter case a grid of points is automatically generated from the information supplied (i.e. points in u-direction and points in v-direction). Note that here the functions are numbered beginning from 1 and the numbering of the basis functions is consecutive, meaning that for the two-dimensional case the basis functions are numbered continuously, first in the u and then in the v direction.

```
function [Ri,idx] = nrbbasisfun(points, nurbs)
% ------------------------------------------------------
% Evaluate non-zero NURBS functions at parametric points.
%
% Input:
% points ... parametric points
% For individual points:
%   NURBS curves:
%   Array of size (number of points= npoints)
%   t (n)=u
```

```
%    NURBS surfaces: Array of size (2, npoints)
%    t (1, n)= u,   t (2, n)= v
% Points on a grid nu*nv
%    2 arrays of size (number of points in u/v directions)
%    {u(n), v(n)}
% nurbs ... NURBS structure
%
% Output:
% Ri(1:npoints,1:nbf)  ... value of basis functions
% where nbf is the number of non-zero basis functions
% idx(1:npnts,1:nbf) ... indexes of non-zero basis functions
%-------------------------------------
```

Example: NURBS curve

```
Ri=nrbbasisfun([0 0.5 1],nurbs)
Ri =1.00000    0.00000    0.00000
     0.29291    0.41418    0.29291
     0.00000    0.00000    1.00000
```

Example: NURBS surface

```
[Ri,idx]= nrbbasisfun([0 0.5 1; 0 0.5 1],nurbs)
Ri =
     1.00000    0.00000    0.00000    0.00000
     0.25000    0.25000    0.25000    0.25000
     0.00000    0.00000    0.00000    1.00000
idx =
     1    2    3    4
     1    2    3    4
     1    2    3    4
```

Notes: The program only computes the shape functions that are not zero for a parametric point. The indexes of the non-zero basis functions are supplied in the array idx.

The next function computes first derivatives of basis functions.

```
function [Ru,Rv,idx] = nrbbasisfunder(points, nurbs)
% -----------------------------------------------------------
% Evaluate non-zero NURBS function derivatives at parametric points.
%
% Input:
% points ... parametric points
% For individual points:
%    NURBS curves:
%    Array of size (number of points= npoints)
%    t (n)=u
```

```
%   NURBS surfaces: Array of size (2, npoints)
%    t (1, n) = u,   t (2, n) = v
% Points on a grid nu*nv
%   2 arrays of size (number of points in u/v directions)
%    {u(n), v(n)}
% nurbs ...  NURBS structure
%
% Output:
% Ru(1:npoints,1:nbf)  ...  first derivative to u of basis functions
% NURBS surfaces only:
% Rv(1:points,1:nbf)  ...  first derivative to v of basis functions
% idx(1:npnts,1:nbf) ... indexes of non-zero basis functions
%-------------------------------------
```

Example, NURBS curve:

```
Ru= nrbbasisfunder([0 0.5 1],nurbs)
Ru =
  -1.41400    1.41400    0.00000
  -1.17165    0.00000    1.17165
   0.00000   -1.41400    1.41400
```

Example, NURBS surface:

```
[Ru,Rv] = nrbbasisfunder([0 0.5 1;0 0.5 1], nurbs)
Ru =
  -1.00000    1.00000    0.00000    0.00000
  -0.50000    0.50000   -0.50000    0.50000
   0.00000    0.00000   -1.00000    1.00000
Rv =
  -1.00000    0.00000    1.00000    0.00000
  -0.50000   -0.50000    0.50000    0.50000
   0.00000   -1.00000    0.00000    1.00000
```

A function for computing the first and second derivatives is supplied only for B-splines. This function requires a call to findspan.

```
function dR = basisfunder(s,p,t,knots,nders)
% --------------------------------------------------------
% Evaluate non-zero B-spline functions and derivatives at parametric points.
%
% Input:
% s ...  knot span (from findspan)
% p ...  order
% t(1:npoints)  ...  parametric points
% knots ...  knot vector
% nders ...  Number of derivatives +1
%
```

```
% Output:
% dR(1:nbf,1,:) ...  B-spline
% dR(1:nbf,2,:) ...  first derivative
% dR(1:nbf,3,:) ...  second derivative
%----------------------------------------
```

Example:

```
s= findspan(2,2,0.5,[0 0 0 1 1 1])
dR = basisfunder(s,2,0.5,[0 0 0 1 1 1],2)
dR =
ans(:,:,1) =
   0.25000  -1.00000   2.00000
ans(:,:,2) =
   0.50000   0.00000  -4.00000
ans(:,:,3) =
   0.25000   1.00000   2.00000
```

Further functions will be explained as required in the later chapters.

5   SUMMARY AND CONCLUSIONS

In this chapter we have introduced the most important aspect of numerical simulation: The basis functions that will be used to describe the problem geometry and to approximate the variation of the unknowns. For the computational efficiency of the simulation it is important to be aware of the computational effort involved in the evaluation of the basis functions and their derivatives, since this has to be done very frequently. The operation count for NURBS functions is explained in detail in

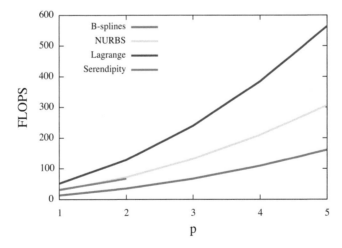

Figure 30   Number of floating point operations (FLOPS) for evaluating two-dimensional basis functions of various orders *p*. Results for Serendipity functions are only shown for up to order 2.

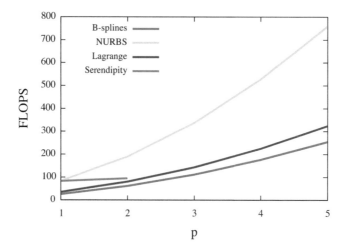

Figure 31   Number of floating point operations (FLOPS) for evaluating the first derivative of various orders *p*. Results for Serendipity functions are only shown for up to order 2.

Table 1   Summary of properties of basis functions.

Basis Function:	Lagrange Serendipity	NURBS
+	easy to implement, fast computation	good control of shape, control of continuity, good refinement strategies order elevation via knot vector
−	not very good control of shape, can exhibit oscillatory behavior, no control over continuity	slightly more difficult to implement, slower computation

[5]. Figure 30 shows the number of floating point operations for the evaluation of a basis function and Figure 31 for its first derivative. NURBS function evaluations are comparable to evaluations of Serendipity functions but the computational effort for evaluating derivatives is larger that for the other basis functions.

In Table 1 we summarize the positive and negative properties of the basis functions introduced here. It obvious that NURBS are superior for describing geometry and for approximating the unknown. The use of the basis functions to describe geometry will be explained next, where the superiority of NURBS will be shown.

BIBLIOGRAPHY

[1]   Carl de Boor. *A practical guide to splines*. Springer, New York, NY, 2001.
[2]   I. Ergatoudis, B.M. Irons, and O.C. Zienkiewicz. Curved isoparametric quadrilateral elements for finite element analysis. *International Journal for Numerical Methods in Engineering*, 4:31–42, 1968.

[3] T. Greville. Numerical procedures for interpolation by spline functions. *J. Soc. Ind. Appl. Math. Ser. B. Numer. Anal.*, 1964.

[4] J.C. Lachat and J.O. Watson. Effective numerical treatment of boundary integral equations: A formulation for three-dimensional elastostatics. *International Journal for Numerical Methods in Engineering*, 10(5):991–1005, 1976.

[5] Benjamin Marussig, Jürgen Zechner, Gernot Beer, and Thomas-Peter Fries. Fast isogeometric boundary element method based on independent field approximation. *Computer Methods in Applied Mechanics and Engineering*, 284(0):458–488, 2015. Isogeometric Analysis Special Issue.

[6] Les Piegl and Wayne Tiller. *The NURBS book (2nd ed.)*. Springer-Verlag, New York, Inc., New York, NY, USA, 1997.

[7] Daniel Rypl and Borek Patzak. Object oriented implementation of the T-spline based isogeometric analysis. *Advances in Engineering Software*, 50:137–149, 2012.

[8] Thomas W. Sederberg, David L. Cardon, G. Thomas Finnigan, Nicholas S. North, Jianmin Zheng, and Tom Lyche. T-spline simplification and local refinement. *ACM Trans. Graph.*, 23(3):276–283, August 2004.

[9] M. Spink, D. Claxton, C. de Falco, and R. Vázquez. The NURBS toolbox. http://octave.sourceforge.net/nurbs/index.html.

Chapter 3

Stage 2: Geometry

Let no man ignorant of geometry enter here
Inscription at the door of Plato's Academy in Athens

One of the applications of basis functions is the approximation of geometrical shapes. This is an essential part of simulation. Therefore, stage 2 of our journey is concerned with the basic concepts of geometrical modeling.

1 COORDINATE SYSTEMS

In geometry, we use a Cartesian orthogonal (right hand) coordinate system. The location of a point in space is given by a position vector $\mathbf{x}$

$$\mathbf{x} = \begin{pmatrix} x \\ y \\ z \end{pmatrix} \tag{1}$$

The other possibilities would be to use cylindrical or polar coordinates. For plane geometry, the z-coordinate is either omitted or set to zero. In keeping with the convention of the NURBS toolkit, we always specify three coordinates, with the z-coordinate set to zero for plane geometry.

1.1 Coordinate transformation

It is sometimes required to transform coordinates from one system to another. Consider a system defined by origin $\mathbf{x}_0$ and orthogonal vectors $\mathbf{v}_1$, $\mathbf{v}_2$ and $\mathbf{v}_3$ as shown in Figure 1.

The transformed original coordinates $\mathbf{x}$, $\bar{\mathbf{x}}$ are given by

$$\bar{\mathbf{x}} = \mathbf{x}_0 + \mathbf{T}^T \mathbf{x} \tag{2}$$

where the transformation matrix is given by

$$\mathbf{T} = (\mathbf{v}_1 \ \mathbf{v}_2 \ \mathbf{v}_3) \tag{3}$$

In the following we define geometrical shapes by mapping from a local coordinate system to the global Cartesian system.

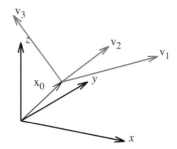

Figure 1 Transformation between 2 Cartesian systems.

2 CURVES

First we deal with curves that have only one parameter direction (ξ or u).

2.1 Mapping with Serendipity/Lagrange basis functions

The mapping from the local parameter space ξ to the global space $\mathbf{x}$ with Serendipity basis functions is given by:

$$\mathbf{x} = \sum_{i=1}^{I} N_i(\xi)\mathbf{x}_i \tag{4}$$

where I is the number of nodes, $N_i(\xi)$ are basis functions introduced earlier and $\mathbf{x}_i$ are the coordinates of node i. The global space can be two- or three-dimensional, i.e. $\mathbf{x}$ can have x, y or x, y, z components.

For Lagrange polynomials we have:

$$\mathbf{x} = \sum_{i=1}^{I} L_i(\xi)\mathbf{x}_i \tag{5}$$

where $L_i(\xi)$ are the Lagrange polynomials introduced earlier. For curves, the Serendipity and Lagrange basis functions are identical, except for the numbering (in the Lagrange functions the nodes are numbered consecutively in the local coordinate direction, whereas in the Serendipity functions the corner nodes are numbered first).

For planar curves the vector tangential to the curve in the direction of the local coordinate ξ can be computed by

$$\mathbf{V}_1 = \frac{\partial \mathbf{x}}{\partial \xi} = \sum_{i=1}^{I} \frac{\partial N_i(\xi)}{\partial \xi}\mathbf{x}_i \quad \text{or} \quad \sum_{i=1}^{I} \frac{\partial L_i(\xi)}{\partial \xi}\mathbf{x}_i \tag{6}$$

The vector normal to the curve can be computed by the vector x-product with a unit vector in the z-direction, $\mathbf{v}_z$:

$$\mathbf{N} = \mathbf{V}_1 \times \mathbf{v}_z = \begin{pmatrix} \dfrac{\partial y}{\partial \xi} \\[2ex] -\dfrac{\partial x}{\partial \xi} \\[2ex] 0 \end{pmatrix} \tag{7}$$

A unit vector in direction normal to the curve is then given by:

$$\mathbf{n} = \frac{\mathbf{N}}{J} \tag{8}$$

where $J = \sqrt{N_x^2 + N_y^2}$ is also known as the *Jacobian* of the mapping.

2.2 Mapping with NURBS

Similarly, the mapping from local parameter space u to $\mathbf{x}$ with NURBS functions is given by:

$$\mathbf{x} = \sum_{i=0}^{I} R_{i,p}(\mathbf{u})\mathbf{x}_i \tag{9}$$

where $I+1$ is the number of control points, $R_{i,p}(\mathbf{u})$ are basis functions introduced earlier and $\mathbf{x}_i$ are coordinates of control point i.

Since Octave does not accept an index 0 we change Equation (9) so that the count starts with 1 and drop the subscript p:

$$\mathbf{x} = \sum_{n=1}^{N} R_n(\mathbf{u})\mathbf{x}_n \tag{10}$$

where N is the number of control points.

For planar curves the vector tangential to the curve in the direction of the local coordinate can be computed by

$$\mathbf{V}_1 = \frac{\partial \mathbf{x}}{\partial \mathbf{u}} = \sum_{i=1}^{I} \frac{\partial R_n(\mathbf{u})}{\partial \mathbf{u}}\mathbf{x}_i \tag{11}$$

The vector normal to the curve is given by:

$$\mathbf{N} = \mathbf{V}_1 \times \mathbf{v}_z = \begin{pmatrix} \dfrac{\partial y}{\partial \mathbf{u}} \\[2ex] -\dfrac{\partial x}{\partial \mathbf{u}} \\[2ex] 0 \end{pmatrix} \tag{12}$$

3 PROGRAMMING

Here we show some additional OCTAVE functions for NURBS curves. We start with explaining the input and output of the functions of the NURBS toolkit, that we will use.

3.1 NURBS toolkit

The first function is to compute the global coordinates at parametric points (u) using Equation (10).

```
function xy = nrbeval(nurbs, tt)
%------------------------------------------------------------
% NURBS toolkit:
% Evaluate NURBS curve at parametric points
%
% Input:
% nurbs ... NURBS structure
% tt (n)... parametric points
%
% Output:
% xy (1:3,n)... global coordinates at parametric points (z=0)
%------------------------------------------------------------
```

The next functions sets up a structure that is required to compute derivatives.

```
function dnurbs = nrbderiv(nurbs)
%------------------------------------------------------------
% NURBS toolkit:
% Sets up the structure for computing the derivatives
%
% Input:
% nurbs ...  NURBS structure
%
% Output:
% dnurbs ...  NURBS structure for derivatives
%------------------------------------------------------------
```

Function *nrbdeval* has the same purpose as *nrbeval* but also computes the first and second derivatives.

```
function [xy,Jac,Hess] = nrbdeval(nurbs,dnurbs, tt)}
%------------------------------------------------------------
% NURBS toolkit:
% Evaluate NURBS curve at parametric points
```

```
% including the first and second derivatives
%
% Input:
% nurbs ...  NURBS structure
% dnurbs ...  NURBS structure for derivatives
% tt (n)...  parametric points
%
% Output:
% xy (1:3,n)...  global coordinates at parametric points
% Jac(1:3,n) ...  first derivatives at parametric points
% Hess(1:3,n) ...  second derivatives at parametric points
%-------------------------------------------------------------
```

3.2   Geometry functions

Here we develop some functions for graphical output, which will be useful later in this book.

Plotting curves

Function *PlotCurv* produces data that allows the plotting of curves.

```
function PlotCurve(coefs,knot)
%------------------------------------
%    Creates data for plotting a planar curve,
%    control points and variation of the Jacobian
%    Input:
%    coefs  ...  weighted control point coords and weights
%    knot   ...  knot vector
%
%    Output to files;
%    "Geometry"  ... contains geometry data
%    "Control"  ... contains control point data
%    "Jacobian"  ... contains Jacobian data
%------------------------------------
fid= fopen("Geometry","w"); fid1= fopen("Control","w");
fid2= fopen("Jacobian","w");
%  Control point coordinates
for n=1:columns(coefs)
  w= coefs(4,n);
  fprintf(fid1,"%8.5f %8.5f \n", coefs(1,n)/w, coefs(2,n)/w);
end
nurbs= nrbmak(coefs,knotu); dnurbs= nrbderiv(nurbs);
%  plot curve and Jacobian
ut= linspace(0,1,40);
```

```
[xyg,vu]= nrbdeval(nurbs,dnurbs,ut);
for nu=1:length(ut)
    Jac= (vu(1,nu)^2+vu(2,nu)^2)^0.5;
    fprintf(fid2,"%8.5f %8.5f  \n", ut(nu),Jac);
    fprintf(fid,"%8.5f %8.5f  \n", xyg(1:2,nu),);
end
fclose(fid);fclose(fid1);fclose(fid2);
endfunction;
```

Calculation of control point locations and weights for circular arc

NURBS basis functions of order $p=2$ are able to exactly represent a circular arc. Here we show how to compute the control point coordinates and weights for a circular arc of radius R and start angle α_s and end angle α_e. If the sustained angle $(\alpha_e - \alpha_s)$ is smaller than $\frac{\pi}{2}$ then we substitute $\alpha_1 = \alpha_s$ and $\alpha_2 = \alpha_e$ and compute 3 control points.

The coordinates and weights of the control points are computed by (see Figure 2):

$$x_1 = x_0 + R \cdot \cos\alpha_1$$
$$y_1 = y_0 + R \cdot \sin\alpha_1 \tag{13}$$
$$\mathrm{w}_1 = 1$$

$$x_2 = x_1 - R \cdot \tan\alpha \cdot \sin\alpha_1$$
$$y_2 = y_1 + R \cdot \tan\alpha \cdot \cos\alpha_1 \tag{14}$$
$$\mathrm{w}_2 = \cos\alpha$$

where $\alpha = (\alpha_2 - \alpha_1)/2$.

$$x_3 = x_0 + R \cdot \cos\alpha_2$$
$$y_3 = y_0 + R \cdot \sin\alpha_2 \tag{15}$$
$$\mathrm{w}_3 = 1.0$$

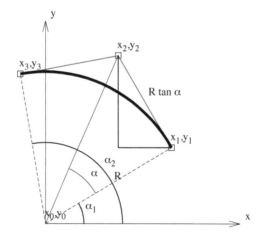

Figure 2 Computation of control point locations and weights for a circular arc.

The knot vector is given by:

$$\Xi = \begin{pmatrix} 0 & 0 & 0 & 1 & 1 & 1 \end{pmatrix} \tag{16}$$

If the sustained angle $(\alpha_e - \alpha_s)$ is larger than $\frac{\pi}{2}$ but smaller than π we set $\alpha_1 = \alpha_s$, $\alpha_2 = \frac{\pi}{2}$ and compute 3 control points using Equations (13) to (15). We then set $\alpha_1 = \frac{\pi}{2}$, $\alpha_2 = \alpha_e$ and compute two additional control points using Equations (14) and (15). The knot vector is given by

$$\Xi = \begin{pmatrix} 0 & 0 & 0 & 0.5 & 0.5 & 1 & 1 & 1 \end{pmatrix} \tag{17}$$

Note: Although the repetition of knots means that the continuity is reduced to C^0, the curve is actually C^1 continuous, due to the location of the control points.

If the sustained angle $(\alpha_e - \alpha_s)$ is larger than π but smaller than $\frac{3\pi}{2}$ we set $\alpha_1 = \alpha_s$, $\alpha_2 = \frac{\pi}{2}$ and compute three control points using Equations (13) to (15). We then set $\alpha_1 = \frac{\pi}{2}$, $\alpha_2 = \pi$ and compute two additional control points using Equations (14) and (15). We then set $\alpha_1 = \pi$, $\alpha_2 = \alpha_e$ and compute two additional control points using Equations (14) and (15). The knot vector is given by

$$\Xi = \begin{pmatrix} 0 & 0 & 0 & 0.3 & 0.3 & 0.6 & 0.6 & 0 & 0 & 0 \end{pmatrix} \tag{18}$$

If the sustained angle $(\alpha_e - \alpha_s)$ is larger than $\frac{3\pi}{2}$ we set $\alpha_1 = \alpha_s$, $\alpha_2 = \frac{\pi}{2}$ and compute 3 control points using Equations (13) to (15). We then set $\alpha_1 = \frac{\pi}{2}$, $\alpha_2 = \pi$ and compute two additional control points using Equations (14) and (15). We then set $\alpha_1 = \pi$, $\alpha_2 = \frac{3\pi}{2}$ and compute two additional control points using Equations (14) and (15). Finally we assign $\alpha_1 = \frac{3\pi}{2}$, $\alpha_2 = \alpha_e$ and compute two additional control points using Equations (14) and (15). The knot vector is given by

$$\Xi = \begin{pmatrix} 0 & 0 & 0 & 0.25 & 0.25 & 0.5 & 0.5 & 0.75 & 0.75 & 0 & 0 & 0 \end{pmatrix} \tag{19}$$

This is converted into two OCTAVE functions as follows: The first function computes the control point coordinates and weights for an arc segment with a sustained angle of $< \frac{\pi}{2}$. The second function uses the first for the computation of control point coordinates, weights and knot vector for an arc with any value of sustained angle.

```
function coefs=Arcseg(x0,R,alfa1,alfa2)
%-----------------------------------------
%  determines control point coords for arc
%  with a sustained angle of  less than or equal to 1.5707 rad
%
%  Input:
%  x0 ...  origin of arc
%  R ... radius
%  alfa1  ...  start angle (radians)
%  alfa2  ...  end angle (radians)
%
```

```
%  Output:
%  coefs  ...  weighted control point coords and weights
%-------------------------------------------
alfa= (alfa2 - alfa1)/2; n=1;
x1= x0(1) + R*cos(alfa1); coefs(1,n)= x1;
x2= x0(2) + R*sin(alfa1); coefs(2,n)= x2;
coefs(3,n)= 0.; coefs(4,n)= 1.0;
n=2; w=cos(alfa);
coefs(1,n)= ( x1 - R*tan(alfa)*sin(alfa1))*w;
coefs(2,n)= ( x2 + R*tan(alfa)*cos(alfa1))*w;
coefs(3,n)= 0.; coefs(4,n)= w;
n=3; coefs(1,n)= x0(1) + R*cos(alfa2);
coefs(2,n)= x0(2) + R*sin(alfa2);
coefs(3,n)= 0.; coefs(4,n)= 1.0;
endfunction;
```

```
function [coefs,knot]=Arc(x0,R,alfas,alfae)
%-------------------------------------------
%  determines parameters and knot vector for a circular arc
%
%  Input:
%  x0 ...  origin of arc
%  R ...  radius
%  alfas  ...  start angle (radians)
%  alfae  ...  end angle (radians)
%
%  Output:
%  coefs  ...  weighted control point coords and weights
%  knot ...  knot vector
%-------------------------------------------
sangle= alfae-alfas; n=0;
if(sangle <= pi/2)
 knot=[0 0 0 1 1 1]; [coefs]= Arcseg(x0,R,alfas,alfae,0);
elseif(sangle > pi/2 && sangle <= pi)
 knot=[0 0 0 0.5 0.5 1 1 1];
 coefs=Arcseg(x0,R,alfas,pi/2);coefss=Arcseg(x0,R,pi/2,alfae);
 i=3; for n=2:3; i=i+1; coefs(1:4,i)=coefss(1:4,n); end
elseif(sangle > pi && sangle <= 3*pi/2)
 knot=[0 0 0 0.3 0.3 0.6 0.6 1 1 1];
 coefs= Arcseg(x0,R,alfas,pi/2); coefss=Arcseg(x0,R,pi/2,pi);
 i=3; for n=2:3; i=i+1; coefs(1:4,i)=coefss(1:4,n); end
 coefss= Arcseg(x0,R,pi,alfae);
 i=5; for n=2:3; i=i+1; coefs(1:4,i)=coefss(1:4,n); end
elseif(sangle > 3*pi/2)
 knot=[0 0 0 0.25 0.25 0.5 0.5 0.75 0.75 1 1 1];
```

```
coefs= Arcseg(x0,R,alfas,pi/2); coefss=Arcseg(x0,R,pi/2,pi);
i=3; for n=2:3; i=i+1; coefs(1:4,i)=coefss(1:4,n); end
coefss= Arcseg(x0,R,pi,3*pi/2);
i=5; for n=2:3; i=i+1; coefs(1:4,i)=coefss(1:4,n); end
coefss= Arcseg(x0,R,3*pi/2,alfae);
i=7; for n=2:3; i=i+1; coefs(1:4,i)=coefss(1:4,n);end
endif
endfunction;
```

3.3   Examples

To better understand the nature of the mapping with different basis functions, we show 3 examples. The first one is a circular arc and serves as a comparison of the approximations by Lagrange polynomials and NURBS. The second example is a practical example of a tunnel section and is meant to demonstrate the flexibility of NURBS to model real shapes. It is also used to examine the effect of the local coordinates of knots on the distribution of the Jacobian. The third example finally demonstrates how NURBS can model discontinuous curves.

3.4   Example 1: Circular arc

This example was chosen to compare the accuracy that can be achieved by an approximation by Lagrange polynomials and NURBS. For the approximation of a circular arc Figure in 3 with Lagrange functions we define the nodal point coordinates as

$$
\mathbf{x}_i = \begin{pmatrix} 1 & 0 \\ \dfrac{\sqrt{2}}{2} & \dfrac{\sqrt{2}}{2} \\ 0 & 1 \end{pmatrix}
\tag{20}
$$

For the approximation by NURBS, the following values for the parameters for *nrbmak* are obtained from function Arc:

```
[coefs,knot]=Arc([0,0],1,0,pi/2)
coefs =
    1.00000    0.70711    0.00000
    0.00000    0.70711    1.00000
    0.00000    0.00000    0.00000
    1.00000    0.70711    1.00000
knot = 0    0    0    1    1    1
```

Figure 3 shows the two different approximations of the arc.

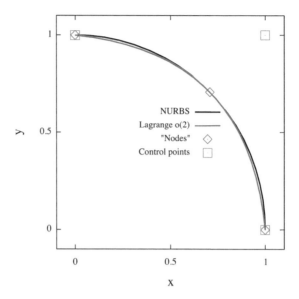

Figure 3 Geometrical description of arc: Comparison of approximation by Lagrange and NURBS basis functions.

To investigate the accuracy of the approximations with the different basis functions we introduce the geometrical error norm as

$$\|\epsilon_g\| = \int_s \frac{|\mathbf{x}^{ap} - \mathbf{x}^{ex}|}{|\mathbf{x}^{ex}|} \cdot ds \tag{21}$$

where $\mathbf{x}^{ap}$ and $\mathbf{x}^{ex}$ are the approximate and exact coordinates on the curve respectively and the integration is along the curve s.

The error in approximating the Length is

$$\|\epsilon_L\| = \frac{L^{ap}}{L^{ex}} \tag{22}$$

where L^{ap} and L^{ex} are the approximate and exact values of length respectively.

A mapping with NURBS exactly represents the arc, i.e. the geometrical error and the error in computing the length is zero. In Figure 4 we plot the error in the approximation of the geometry and the length as a function of the number of nodal points (using order elevation) for the approximation with Lagrange functions.

In Figure 5, we investigate the variation of the Jacobian for the approximation with Lagrange polynomials. We see that the order of the function influences the variation of the Jacobian. For the higher order functions the Jacobian tends towards a constant value. The variation of the Jacobian for the NURBS representation is shown on the right in Figure 5. It can be seen that the Jacobian is not constant, which is somewhat surprising, since the arc is exactly represented. However, as will be shown later the variation of the Jacobian is closely related to the knot vector.

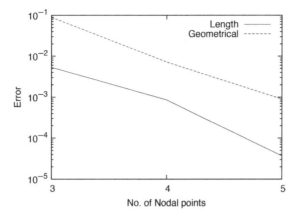

Figure 4   Diagram of errors in approximation of the geometry of the arc and its length by Lagrange functions with different number of nodes.

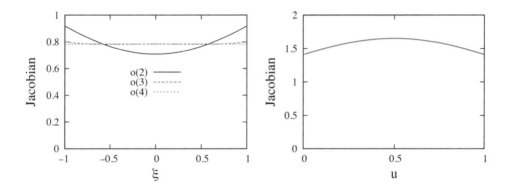

Figure 5   Left: Variation of Jacobian for the approximation with Lagrange function of different orders and right: Variation of Jacobian for approximation by NURBS.

3.5   Example 2: Horseshoe tunnel

A practical application is the representation of the geometry of a horseshoe tunnel[1]. The cross-section of the tunnel is usually specified by engineers using arcs as shown in Figure 6. The information provided includes centers and radii of the arcs. Arc centers and start/end angles are chosen in such a way so that there is a smooth transition from one arc to the other. For tunnels it is important to have such smooth cross-sections, as otherwise we would have stress concentrations. Indeed, the shape of the tunnel is chosen to minimize the magnitude of stresses in the lining. This represents an ideal

---

[1]This is a typical cross-section for a tunnel using the New Austrian Tunneling Method or NATM.

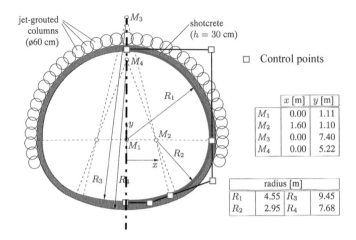

Figure 6 Definition of the geometry of a tunnel.

application of NURBS basis functions, since the design geometry can be represented exactly.

Using the Arcseg function introduced previously we determine the parameters and the knot vector with the function NATM.

```
function [coefs,knot]=NATM
%-------------------------------------------
%   determines parameters and knot vector
%   for an NATM tunnel
%
%   Output:
%   coefs  ...  weighted control point coords and weights
%   knot ... knot vector
%-------------------------------------------
 knot=[0 0 0 0.5 0.5 0.8 0.8 1 1 1];
 coefss= Arcseg([0,1.11],4.55,0,pi/2);
 coefs(1:4,3)=coefss(1:4,1); coefs(1:4,2)=coefss(1:4,2);
 coefs(1:4,1)=coefss(1:4,3);
 coefss= Arcseg([1.6,1.1],2.95,4.939,2*pi);
 coefs(1:4,4)=coefss(1:4,2); coefs(1:4,5)=coefss(1:4,1);
 coefss= Arcseg([0,7.4],9.45,3*pi/2,4.939);
 coefs(1:4,6)=coefss(1:4,2); coefs(1:4,7)=coefss(1:4,1);
endfunction;
```

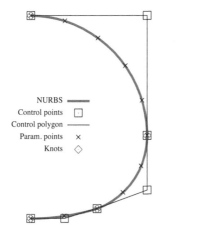

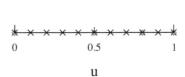

NURBS	━━━			
Control points	☐			
Control polygon	────			
Param. points	×			
Knots	◇			

Figure 7 Mapping of shape of tunnel from the local u coordinate system (right) to global x, y coordinates (left). Knots are depicted by red diamonds.

The result is:

```
coefs =
    0.00000    3.21734    4.55000    3.56047    2.26280    1.06845   -0.00000
    5.66000    4.00222    1.11000   -0.97597   -1.77458   -2.03685   -2.05000
    0.00000    0.00000    0.00000    0.00000    0.00000    0.00000    0.00000
    1.00000    0.70711    1.00000    0.78252    1.00000    0.99359    1.00000
knots = 0    0    0    0.5    0.5    0.8    0.8    1    1    1
```

The mapping of the shape of the tunnel is shown in Figure 7. Because of repetition of knots in the knot vector there is only a C^0 continuity guaranteed at this location, but actually the control polygon indicates a C^1 continuity. This is because the locations of the arcs have been chosen such that there is a unique tangent where arcs meet. However, as can be seen in Figure 9 the Jacobian is discontinuous at the knots.

Next we investigate the effect of the location of the knots by shifting the location of one from 0.5 to 0.4 in the knot vector, i.e. the knot vector is changed to $\Xi = (0\ 0\ 0\ 0.4\ 0.4\ 0.8\ 0.8\ 1\ 1\ 1)$.

It can be seen in Figure 8 that this has no effect on the shape. The effect is only on the distribution of the parametric points, which are wider spaced out in the top part now. However, this also affects the distribution of the Jacobian (see Figure 9).

3.6   Example 3: Plate with hole

The third example was chosen to show how easy it is to control the continuity with NURBS. Figure 10 shows an example of a quarter of a plate with a circular hole. We choose the order of the basis functions to be $p = 2$. For this case we need to have double entries in the Knot vector at each corner so that we have a C^0 continuity there.

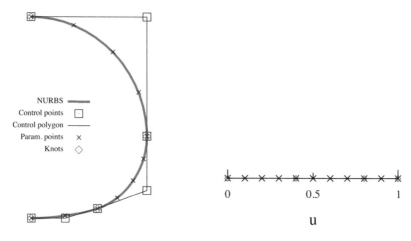

Figure 8 Mapping of shape of tunnel from the local u coordinate system (right) to global x, y coordinates (left) with changed knot value.

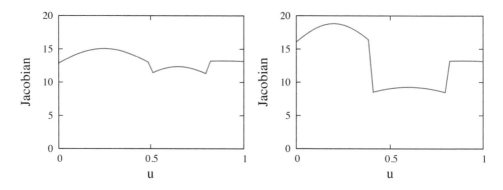

Figure 9 Variation of Jacobian for tunnel with (left) original and (right) changed knot value.

The quarter plate with a hole is exactly represented with this NURBS definition. The input data for function *Plotcurve* are:

```
coefs=
0.5 0.353 1.0 1.0 1.0 0.5 0.0 0.0 0.0 0.25 0.5
0.0 0.353 0.5 0.75 1.0 1.0 1.0 0.5 0.0 0.0 0.0
0 0 0 0 0 0 0 0 0 0 0
1 0.707 1 1 1 1 1 1 1 1 1
knot= 0 0 0 1 1 2 2 3 3 4 4 5 5 5
```

4 SURFACES

Here we deal with surfaces with two coordinates in parameter space (ξ, η or u, v).

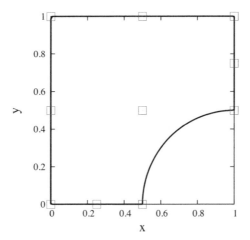

Figure 10 Definition of a quarter plate with a circular hole showing control points.

4.1 Mapping with Serendipity/Lagrange basis functions

The coordinates **x** at location ξ, η are computed, using Serendipity functions:

$$\mathbf{x} = \sum_{i=1}^{I} N_i(\xi, \eta)\mathbf{x}_i \tag{23}$$

and by Lagrange polynomials

$$\mathbf{x} = \sum_{i=1}^{I} \sum_{j=1}^{J} L_{ij}(\xi, \eta)\mathbf{x}_{ij} \tag{24}$$

where **x** is a space vector (i.e. has 3 components). Note that for Serendipity functions the nodes are numbered consecutively, whereas Lagrange polynomials have two indices for the nodes.

We can compute two vectors tangential to the surface (one in the ξ the other in the η direction as

$$\mathbf{V}_1 = \frac{\partial \mathbf{x}}{\partial \xi} = \sum_{i=1}^{I} \frac{\partial N_i(\xi, \eta)}{\partial \xi} \cdot \mathbf{x}_i \quad \text{or} \quad \mathbf{V}_1 = \sum_{i=1}^{I} \sum_{j=1}^{J} \frac{\partial L_{ij}(\xi, \eta)}{\partial \xi} \mathbf{x}_{ij} \tag{25}$$

$$\mathbf{V}_2 = \frac{\partial \mathbf{x}}{\partial \eta} = \sum_{i=1}^{I} \frac{\partial N_i(\xi, \eta)}{\partial \eta} \cdot \mathbf{x}_i \quad \text{or} \quad \mathbf{V}_2 = \sum_{i=1}^{I} \sum_{j=1}^{J} \frac{\partial L_{ij}(\xi, \eta)}{\partial \eta} \mathbf{x}_{ij} \tag{26}$$

The derivatives of the Lagrange polynomials are given by:

$$\frac{\partial L_{ij}(\xi, \eta)}{\partial \xi} = \frac{\partial L_i(\xi)}{\partial \xi} \cdot L_j(\eta) \tag{27}$$

$$\frac{\partial L_{ij}(\xi, \eta)}{\partial \eta} = L_i(\xi) \cdot \frac{\partial L_j(\eta)}{\partial \eta} \tag{28}$$

A vector normal to the surface may be computed by

$$\mathbf{N} = \mathbf{V}_1 \times \mathbf{V}_2 = \begin{pmatrix} \dfrac{\partial y}{\partial \xi} \cdot \dfrac{\partial z}{\partial \eta} - \dfrac{\partial y}{\partial \eta} \cdot \dfrac{\partial z}{\partial \xi} \\[2ex] \dfrac{\partial z}{\partial \xi} \cdot \dfrac{\partial x}{\partial \eta} - \dfrac{\partial z}{\partial \eta} \cdot \dfrac{\partial x}{\partial \xi} \\[2ex] \dfrac{\partial x}{\partial \xi} \cdot \dfrac{\partial y}{\partial \eta} - \dfrac{\partial x}{\partial \eta} \cdot \dfrac{\partial y}{\partial \xi} \end{pmatrix} \tag{29}$$

A unit normal to the surface is given by:

$$\mathbf{n} = \frac{\mathbf{N}}{J} \tag{30}$$

where $J = \sqrt{N_x^2 + N_y^2 + N_z^2}$ is the *Jacobian* of the mapping.

4.2 Mapping with NURBS basis functions

The coordinates $\mathbf{x}$ at location u, v on the NURBS surface are computed by:

$$\mathbf{x} = \sum_{i=0}^{I} \sum_{j=0}^{J} R_{i,j}^{p,q}(u, v)\mathbf{x}_{ij} \tag{31}$$

where $R_{i,j}^{p,q}(u, v)$ are the NURBS basis functions introduced earlier and $\mathbf{x}_{ij}$ are the coordinates of control points.

For the implementation it is convenient to continuously number the control points. Furthermore, since OCTAVE does not accept a zero index, we change the notation so the sum starts from 1 instead of zero.

We re-define the NURBS functions as

$$R_n = R_{i(n),j(n)}^{p,q} \tag{32}$$

and the control points as

$$\mathbf{x}_n = \mathbf{x}_{i(n),j(n)} \tag{33}$$

where $i(n)$ and $j(n)$ are the indices associated with the *n*-th control point.

With this new notation Equation 31 can be rewritten as

$$\mathbf{x} = \sum_{n=1}^{N} R_n(u, v) \mathbf{x}_n \tag{34}$$

with $N = (I + 1) \cdot (J + 1)$

The vectors tangential to the surface are given by

$$\mathbf{V}_1 = \frac{\partial \mathbf{x}}{\partial u} = \sum_{n=1}^{N} \frac{\partial R_n(u, v)}{\partial u} \cdot \mathbf{x}_i$$

$$\mathbf{V}_2 = \frac{\partial \mathbf{x}}{\partial v} = \sum_{n=1}^{N} \frac{\partial R_n(u, v)}{\partial v} \cdot \mathbf{x}_i \tag{35}$$

4.3 Programming

Here we first introduce the functions in the NURBS toolkit that computes the global coordinates and the derivatives at a point with the local coordinates u, v. This function is the same as introduced for curves previously but with a different parameter list.

```
function [x,Jac,Hess] = nrbdeval(nurbs,dnurbs,tt)
%-----------------------------------------------------------
% Computes global coordinates and derivatives at parametric
% points tt
%
% Input:
%  nurbs ...  NURBS structure
%  dnurbs  ... NURBS structure for derivatives
%  tt ...     parametric points:
%     For np individual points:
%         Array of size (2,np)
%         tt (1, n)= u,   tt (2, n)= v
%     For points on a grid nu*nv
%         2 arrays of size (nu/nv)
%         u(n),v(n)
%
% Output:
% x ...   global coordinates at parametric points
%     For individual points:
%      Array of size (3,np)
%         x= x(1,n), y= x(2,n), z= x(3,n)
%      For points on a grid nu*nv:
%         Array of size (3, nu,nv)
%         x= x(1,i,j), y= x(2,i,j), z= x(3,i,j)
%
```

```
% Jac{1}, Jac{2} ...  first derivatives of x
%     For individual points:
%         2 arrays of size (3,n)
%         Jac{1}(1,n)=dx/du(n), Jac{1}(2,n)=dy/du(n),
%         Jac{1}(3,n)=dz/du(n)
%         Jac{2}(1,n)=dx/dv(n), Jac{2}(2,n)=dy/dv(n),
%         Jac{2}(3,n)=dz/dv(n)
%     For points on a grid nu*nv
%         2 arrays of size (3,nu,nv)
%         Jac{1}(1,i,j)=dx/du(i,j), Jac{1}(2,i,j)=dy/du(i,j),
%         Jac{1}(3,i,j)=dz/du(i,j)
%         Jac{2}(1,i,j)=dx/dv(i,j), Jac{2}(2,i,j)=dy/dv(i,j),
%         Jac{2}(3,i,j)=dz/dv(i,j)
%
% Hess{1}, Hess{2} ...  second derivatives of x
%     Same as for Jac
%-------------------------------------------------------------
```

A function for plotting surfaces, control points and the variation of the Jacobian is shown next.

```
function PlotSurface(coefs,knotu,knotv)
%------------------------------------
%     Creates data for plotting a surface
%     Input:
%     coefs ... homogenized coordinates and weights
%     knotu  ...  knot vector in u-direction
%     knotv  ...  knot vector in v-direction
%
%     Output to file
%     "Geometry"  ... contains geometry data
%     "Control"   ... contains control points data
%     "Jacobian"  ... contains Jacobian data
%------------------------------------
nurbs= nrbmak(coefs,{knotu,knotv}); ncu= nurbs.number(1);
ncv= nurbs.number(2);
%  plot control points
nctrl=0;
for icv=1:ncv
 for icu= 1:ncu
  w= coefs(4,icu,icv); nctrl= nctrl+1;
  fprintf(fid1,"%8.5f %8.5f %8.5f %d \n", coefs(1:3,icu,icv)/w,nctrl);
 end
end
ut= linspace(0,1,20); vt= linspace(0,1,20);
% plot surface and Jacobian
dnurbs= nrbderiv(nurbs);
[xyg,vuv]= nrbdeval(nurbs,dnurbs,{ut,vt});
for nv=1:length(vt)
```

```
for nu=1:length(ut)
    fprintf(fid,"%8.5f %8.5f %8.5f \n", xyg(1:3,nu,nv));
    norm= vexp(vuv{1}(:,nu,nv),vuv{2}(:,nu,nv));
    Jac= (norm(1)^2+norm(2)^2+norm(3)^2)^0.5;
    fprintf(fid2,"%8.5f %8.5f %8.5f \n",ut(nu),vt(nv),Jac);
  end
 fprintf(fid,"\n"); fprintf(fid2,"\n")
end
fprintf(fid,"\n"); fprintf(fid,"\n");
fclose(fid);fclose(fid1);fclose(fid2);
endfunction;
```

5 SURFACE OF REVOLUTION

Surfaces may be created by revolving a curve (Generatrix) around an axis of revolution (see Figure 11).

To compute the control point coordinates of a surface of revolution, the following procedure is used:

- Establish a local coordinate system by taking the vector x-product between the vector defining the axis of revolution and a vector normal to it pointing to the base of the generatrix (see Figure 11, right).
- In this local system compute the coordinates of the control points corresponding to the start and end angle by rotating the control points of the generatrix around the axis of revolution using the previously introduced function Arc.
- Transform the coordinates back to the global system.

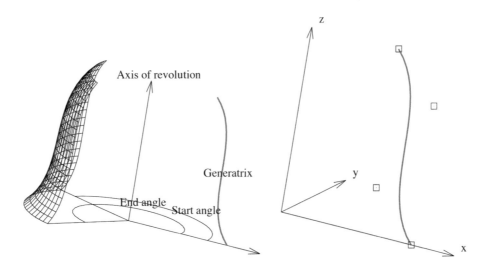

Figure 11   Definition of a surface of revolution and local coordinate system.

The function Surfrev below creates the data for plotting a surface of revolution.

```
function [coefs,knotu,knotv]= Surfrev(axis,gtrix,alfas,alfae)
%------------------------------------
%    Creates data for surface of revolution
%  Input:
%  axis  ...  vector defining axis of revolution
%  gtrix  ...  NURBS structure of generatrix curve
%  alfas  ...  start angle
%  alfae  ...  end angle
%
%  Output:
%  coefs  ...  control point coords and weights
%  knotu,knotv  ...  knot vectors in u- and v-direction
%------------------------------------
coefs= gtrix.coefs; ncu= gtrix.number; knotv= gtrix.knots;
% control point locations on generatrix
for n=1:ncu
 w= coefs(4,1); xy(1:3,n)= coefs(1:3,n)/w;
end
%  local axes and transformation matrix
v1(1:3)= xy(1:3,1); v3= axis; v2= vexp(v3,v1);
for j=1:3
 T0(j,1)= v1(j); T0(j,2)= v2(j); T0(j,3)= v3(j);
end
x0(1:2)=0.;
for n=1:ncu
 x(1:3,1)= xy(1:3,n);  xt= T0*x; R= sqrt(xt(1)^2+xt(2)^2);
 [coefss,knot]=Arc(x0,R,alfas,alfae);
 for i=1:columns(coefss)
  win= coefss(4,i); xt(1:2,1)= coefss(1:2,i)/win; xt(3,1)= xt(3);
  x= transpose(T0)*xt; coefs(1:3,i,n)= x(1:3)*win; coefs(4,i,n)= win;
 end
end
knotu=knot;
endfunction;
```

As with curves we present several examples in order to better understand how NURBS can be effectively used to model surfaces. We start with two surfaces of revolution that can be exactly represented by NURBS.

5.1 Example 1: Cylindrical surface

The function Cylinder will take a straight line as the generatrix and rotate this around the axis of revolution for a sustained angle of 2π.

```
function Cylinder
%------------------------------
%    Plots a cylindrical surface
```

```
%---------------------------------
coefs= [1 1;0 0;0 3;1 1]; knots= [0 0 1 1];
gtrix= nrbmak(coefs,knots);
axis= [0,0,1]; alfas=0; alfae=2*pi;
[coefs,knotu,knotv]= Surfrev(axis,gtrix,alfas,alfae);
PlotSurface(coefs,knotu,knotv)
endfunction;
```

The plot in Figure 12 is obtained.

5.2 Example 2: Spherical surface

This example is used to show the differences of defining surfaces with Serendipity/Lagrange functions and NURBS and the errors introduced by the former. As mentioned before NURBS are able to exactly represent the geometry. The geometry definitions with the first two functions are shown in Figure 13. To model a quarter of a sphere it is necessary to place 3 nodal points at the same location. The difference between the discretizations is that Lagrange polynomials have an additional nodal point controlling the geometry. Therefore a better approximation of the geometry is expected.

Next we introduce the geometrical error norm as

$$\|\epsilon_g\| = \int_S \frac{|\mathbf{x}^{ap} - \mathbf{x}^{ex}|}{|\mathbf{x}^{ex}|} \cdot dS \tag{36}$$

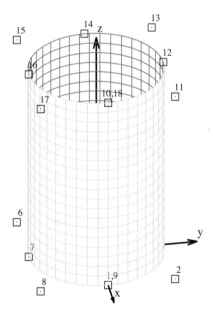

Figure 12 Definition of cylinder with NURBS.

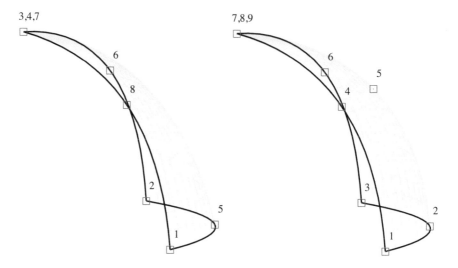

Figure 13 Mapping of spherical surface with (left) Serendipity functions and (right) with Lagrange polynomials.

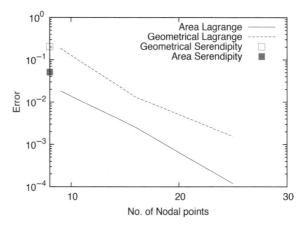

Figure 14 Norm of the error in approximating the geometry for mapping with Lagrange polynomials and Serendipity basis functions.

where $\mathbf{x}^{ap}$ and $\mathbf{x}^{ex}$ are the approximate and exact coordinates on the curve respectively and the integration is along the surface S.

The error in approximating the surface area is

$$\|\epsilon_A\| = \frac{A^{ap}}{A^{ex}} \qquad (37)$$

where A^{ap} is the approximate area and A^{ex} is the exact area.

For the refinement with Lagrange polynomials order elevation is used. The error in the approximation depending on the number of nodal points is shown in Figure 14.

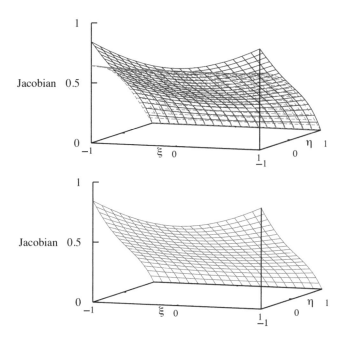

Figure 15 Variation of the Jacobian for approximation with top Lagrange functions and order 2 (black) and 3 (red) and bottom Serendipity functions.

For the approximation with Serendipity functions only the results for 8 nodes (i.e. order 2) are shown.

It can be seen that with a low number of nodal points the errors are significant. The variation of the Jacobian is shown in Figure 15. It can be seen that the Jacobian tends to zero at the point where three nodes have been placed at the same location. It will be shown later that such *degenerate surfaces* can be used in numerical simulation without detriment. A similar effect as for the 2D case can be observed, i.e. that the variation of the Jacobian becomes more constant as the order of the Lagrange functions is increased.

Next we use NURBS to define the geometry. The resulting mapping is shown in Figure 16. Similar to the previous case the last 3 control points have been placed at identical locations. The variation of the Jacobian is plotted in Figure 17.

It can be seen that the Jacobian tends to zero as $v = 1$ (top of sphere) is approached. The error in the approximation of the geometry and area is zero, i.e. the surface is exactly represented.

5.3   Example 3: Bell shaped surface

The final example is designed to demonstrate the ease in which surfaces can be changed by changing just one parameter. It is a Bell shaped surface of revolution, using a generatrix of order 3 (Figure 18). A function that plots this surface is introduced next.

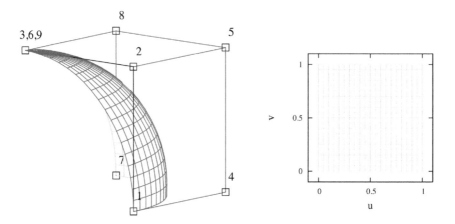

Figure 16 Mapping of spherical surface with NURBS from local u, v to global x, y, z coordinate system.

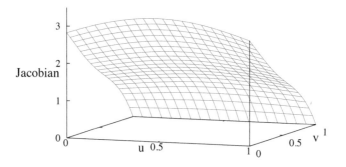

Figure 17 Variation of the Jacobian for spherical surface.

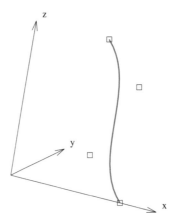

Figure 18 Generatrix of a bell shaped surface.

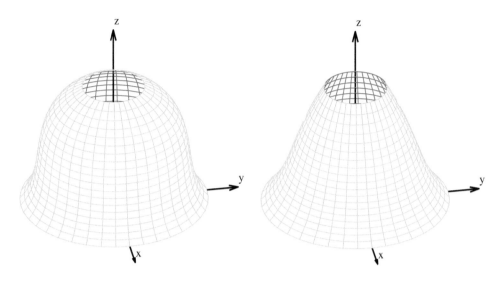

Figure 19 The beauty of NURBS: Generated surface and modified surface changing only the value of the weight at one control point of the generatrix.

```
function Bell
%-------------------------------
%    Creates a bell shaped surface of revolution
%-------------------------------
weigth=1.0; % changed to 0.707
cntrl= [0.75 0 0 1; 0.5 0 0.25 1; 0.75 0 0.75 weight; 0.25 0 1 1];
knots= [0 0 0 0 1 1 1 1];
for n=1:rows(cntrl)
 w= cntrl(n,4); coefs(1:3,n)= cntrl(n,1:3)*w; coefs(4,n)= w;
end
gtrix= nrbmak(coefs,knots);
axis= [0,0,1]; alfas=0; alfae=2*pi;
[coefs,knotu,knotv]= Surfrev(axis,gtrix,alfas,alfae);
PlotSurface(coefs,knotu,knotv)
endfunction;
```

The generated surface is shown in Figure 19, where it is also shown how easily the shape of the surface can be modified by changing only the weight of the penultimate control point of the generatrix from 1.0 to 0.707.

6 LOFTED SURFACES

Lofted surfaces are also generated using an generatrix curve but instead of rotating it is translated along a line. The line indicates the direction and the distance of lofting.

The lofting procedure is as follows:

- Define the generatrix in a local x-y coordinate system
- Define the lofting direction and length
- Compute the distance d of the lofting in a direction normal to the local x-y plane
- Compute the locations of the second set of control points z=d in the local coordinate system
- Transform back to the global system

A function for determining the control points and knot vectors of a lofted surface is shown next.

```
function [coefs,knotu,knotv]= Surftrans(axis,gtrix)
%-------------------------------------
%    Creates data for surface with lofting
%  Input:
%  axis  ... vector defining lofting direction and extent
%  gtrix  ... NURBS structure of generatrix curve
%
%  Output:
%  coefs  ... control point coords and weights
%  knotu,knotv  ... knot vectors in u- and v-direction
%-------------------------------------
[v3,loft] = normalize(axis)
v1= [1 0 0]; v2= vexp(v1,v3);
for j=1:3
 T(j,1)= v1(j);T(j,2)= v2(j);T(j,3)= v3(j);
end
coefsg= gtrix.coefs; ncu= gtrix.number; knot= gtrix.knots;
%    unlofted control points
n=1;
for i=1:ncu
 w= coefsg(4,i); x(1:2,1)= coefsg(1:2,i)/w;
 x(3,1)= 0; xt= T*x;
 coefs(1:3,i,n)= xt(1:3)*w; coefs(4,i,n)= w;
end
%  lofted control points
n=2;
for i=1:ncu
 w= coefsg(4,i); x(1:2,1)= coefsg(1:2,i)/w;
 x(3,1)= loft; xt= T*x;
 coefs(1:3,i,n)= xt(1:3)*w; coefs(4,i,n)= w;
end
knotu=knot;knotv= [0 0 1 1]
endfunction;
```

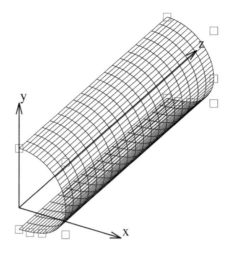

Figure 20 Half of NATM tunnel created by lofting.

We apply the procedure to create half a NATM tunnel surface by lofting the generatrix defining the tunnel along the line (0 25 0) and produce data for plotting.

```
function Tunnel
%----------------------------------------
%  gerates data for plotting a lofted NATM tunnel
%----------------------------------------
axis=[ 0 25 0];
[coefs,knot]=NATM;
gtrix= nrbmak(coefs,knot)
[coefs,knotu,knotv]= Surftrans(axis,gtrix)
PlotSurface(coefs,knotu,knotv)
endfunction;
```

The resulting plot is shown in Figure 20.

7 NURBS SURFACES WITH CUTOUTS

In Figure 21 a cylindrical surface with a cutout is shown. Although it is possible to describe this geometry with the NURBS technology already presented, it becomes rather cumbersome to determine the control point locations especially if the shape of the cutout becomes more complex.

There are two alternative possibilities for the geometrical description (see Figure 22). The first one involves the deletion of the area cutoff by a trimming curve, the second involves generating a T-spline mesh. The first approach is used by CAD programs for the display of the geometry, but is not analysis suitable. The second

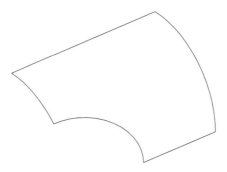

Figure 21   Cylindrical surface with cutout.

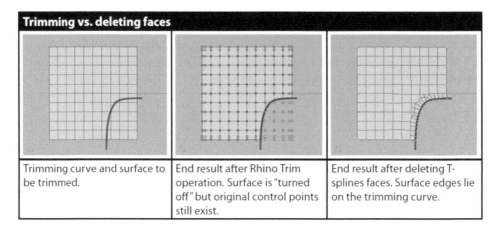

Trimming vs. deleting faces		
Trimming curve and surface to be trimmed.	End result after Rhino Trim operation. Surface is "turned off" but original control points still exist.	End result after deleting T-splines faces. Surface edges lie on the trimming curve.

Figure 22   Two methods for describing a NURBS surface with a cut-out. Image taken from Tsplines 3 Manual ([6]).

approach leads to an analysis suitable geometry description. In the following, ways are shown to make the geometry description by the first approach analysis suitable.

7.1   Analysis suitable trimmed NURBS surfaces

We are aware of only three papers that have addressed the problem of using trimmed CAD data for analysis. The first two papers ([3] and [4]) propose to generate a regular grid of elements, that are defined by knot spans. A searching algorithm is employed, that allows to determine how the elements are transected by trimming curves. The method is very general and can deal with extreme trimming cases, such as multiple holes and cases where trimming curves are very close to each other. However, the implementation of the method is not trivial. The third paper [5] deals with the application of trimming to shell surfaces and with the specification of local loading. The method uses a reconstruction of knot spans and control points and is also generally

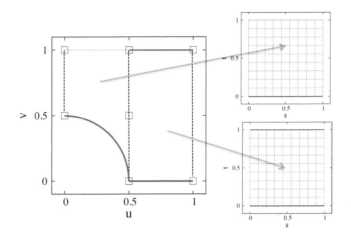

Figure 23 Explanation of mapping trimming curves from u, v to s, t coordinate system. The trimming curves are color coded so their mapping to the s, t coordinate system can be seen.

applicable but not so easy to implement. Here we propose a simple alternative that was published recently in [2].

We start with the definition of the boundary of the cut out in the local parameter space of the quarter cylinder of Figure 21. The trimming curves are shown with their control points in Figure 23. Next we divide the surface in two parts, each delimited by a top and bottom trimming curve and then map each onto a unit square parameter space s, t.

The trimming curves map as straight lines in the s, t coordinate space as shown in Figure 23. For the *nc*-th trimming curve the mapping from the s, t to the u, v parameter space is:

$$u_{nc}(s) = \sum_{n=1}^{N_{nc}} R_{n,nc}(s) \cdot u_{n,nc}; \quad v_n(s) = \sum_{n=1}^{N_{nc}} R_{i,nc}(s) \cdot v_{n,nc} \tag{38}$$

where $R_{n,nc}(s)$ are the NURBS basis functions, N_{nc} is its number of control points and $u_{n,nc}$, $v_{n,nc}$ are the u, v coordinates of control points for curve *nc*.

Next we introduce linear interpolation functions between the curves as

$$N_1(t) = 1 - t$$
$$N_2(t) = t \tag{39}$$

The mapping of points on the trimmed surface from the s, t to the u, v coordinate system is then given by:

$$u = N_1(t) \cdot u_b(s) + N_2(t) \cdot u_t(s)$$
$$v = N_1(t) \cdot v_b(s) + N_2(t) \cdot v_t(s) \tag{40}$$

where the subscripts b and t have been substituted for nc indicating the bottom and top trimming curves respectively.

The Jacobi matrix for the mapping from s, t to u, v coordinates is given by:

$$J_{uv} = \begin{pmatrix} \dfrac{\partial u}{\partial s} & \dfrac{\partial u}{\partial t} \\[2mm] \dfrac{\partial v}{\partial s} & \dfrac{\partial v}{\partial t} \end{pmatrix} \tag{41}$$

The derivatives are given by:

$$\frac{\partial u}{\partial s} = N_1(t) \cdot \frac{\partial u_b(s)}{\partial s} + N_2(t) \cdot \frac{\partial u_t(s)}{\partial s}$$

$$\frac{\partial v}{\partial s} = N_1(t) \cdot \frac{\partial v_b(s)}{\partial s} + N_2(t) \cdot \frac{\partial v_t(s)}{\partial s}$$

$$\frac{\partial u}{\partial t} = -1 \cdot u_b(s) + 1 \cdot u_t(s) \tag{42}$$

$$\frac{\partial v}{\partial t} = -1 \cdot v_b(s) + 1 \cdot v_t(s)$$

where

$$\frac{\partial u_{nc}(s)}{\partial s} = \sum_{n=1}^{N_{nc}} \frac{\partial R_{n,nc}(s)}{\partial s} u_{n,nc} \tag{43}$$

$$\frac{\partial v_{nc}(s)}{\partial s} = \sum_{n=1}^{N_{nc}} \frac{\partial R_{n,nc}(s)}{\partial s} v_{n,nc} \tag{44}$$

The Jacobian is given by

$$J_{uv} = |J_{uv}| \tag{45}$$

The second mapping is from the u, v to the x, y, z coordinate system using Equation (34). The final Jacobian is $J = J_{uv} \cdot J_{xyz}$ where J_{xyz} is the Jacobian of the mapping from u, v to x, y, z coordinates. As can be seen the method, although very easy to implement, is only applicable to cases where the trimmed surface can be defined by opposing trimming curves.

A function that maps points from the s, t to the u, v coordinate system is shown below.

```
function [uv,Jac,i,nc]= Map(Knotg,Coefs,ss,tt,i,nc)
%-------------------------------------
%    Maps points on trimmed surface from s,t to u,v
%    coordinate system
%    Input:
```

```
%    Knotg   ... array containing knot vectors of trimming  curves
%    Coefs  ...    array containing control points of trimming curves
%    ss  ... vector with s-coordinates
%    tt  ... vector with t-coordinates
%    i ... counter for knot values
%    nc ... counter for Coefs
%    Output:
%    uv     ...   u,v coordinates of points
%    Jac    ...   Jacobian
%    i ... updated counter for knot values
%    nc ... updated counter for Coefs
%-----------------------------------
for ncrv=1:2
 clear knotu; clear coefs;
%  read knots and coefs
 [knotu,coefs,i,nc]= Get_infoc(Knotg,Coefs,i,nc);
%   mapping of trimming curves from s,t coordinates
 nurbs= nrbmak(coefs,knotu); dnurbs= nrbderiv(nurbs);
 [uv,vuv]= nrbdeval(nurbs,dnurbs,ss);
 uvn(:,:,ncrv)= uv(:,:); vuvn(:,:,ncrv)=vuv(:,:);
end
%   mapping of points on trimmed surface from s,t to u,v
np=0;
for nt=1:length(tt)
 for ns=1:length(ss)
  N1=1-tt(nt); N2=tt(nt);
% compute u,v values
  np= np + 1; uv(1,np)= N1*uvn(1,ns,1) + N2*uvn(1,ns,2);
  uv(2,np)= N1*uvn(2,ns,1) + N2*uvn(2,ns,2);
%   compute Jacobian
  v1(1)= N1*vuvn(1,ns,1) + N2*vuvn(1,ns,2);
  v1(2)= N1*vuvn(2,ns,1) + N2*vuvn(2,ns,2); v1(3)= 0;
  v2(1)= uvn(1,ns,2) - uvn(1,ns,1);
  v2(2)= uvn(2,ns,2) - uvn(2,ns,1); v2(3)= 0;
  norm= vexp(v1,v2); Jac(np)= norm(3);
 end
end
endfunction;
```

The function uses *Get_infoc* that extracts information about trimming curves.

```
function [knotu,coefs,i,nc]= Get_infoc(Knotg,Coefs,i,nc)
%-----------------------------------
%    Extracts knot vector and parameters for curve
%    Input:
%    Knotg  ... array containint knot vectors
%    Coefs  ... array containing parameters
%    i  ... counter for knot vector
```

```
%    nc ... counter for parameters
%    Output:
%    knotu ...  extracted knot vector
%    coefs ...  extarcted parameters
%    i  ... updated counter for knot vector
%    nc ... updated counter for parameters
%-----------------------------------
i=i+1; ncu= Knotg(i); i=i+1; pu=Knotg(i);
for n=1:ncu+pu+1
   i=i+1; knotu(n)= Knotg(i);
end
for nu=1:ncu
   nc=nc+1;
   w= Coefs(4,nc); coefs(1:3,nu)= Coefs(1:3,nc)*w; coefs(4,nu)= w;
end
endfunction;
```

Finally function *PlotTrimSurface* is presented that can be used to plot a trimmed surface.

```
function PlotTrimSurface(coefs,knotu,knotv,Knotr,Coefstr,ntrims)
%-----------------------------------
%    Creates data for plotting  a trimmed surface
%    Input:
%    coefs ... parameters for surface
%    knotu,knotv ... knot vectors of surface
%    Knotr ... array containing knot vectors of triming curves
%    Coefstr ... array containing paramters of trimming curves
%    ntrims ... number of trimmed surfaces
%    Output to file
%    "Geometry" ... contains geoemetry data in x,y,z coords
%    "Jacobian" ... contains Jacobian data
%-----------------------------------
nss=15; ntt=15; ss= linspace(0,1,ntt); tt= linspace(0,1,nss);
fid= fopen("Geometry","w");fid2= fopen("Jacobian","w"); i=1; nc=0;
for nsurf=1:ntrims
% 1. map from s,t coordinates
 [uv,Jacuv,i,nc]= Map(Knotg,Coefs,ss,tt,i,nc);
% 2. map from u,v to x,y,z coordinates
 nurbs= nrbmak(coefs,{knotu,knotv}); dnurbs= nrbderiv(nurbs);
 [xyg,vuv]= nrbdeval(nurbs,dnurbs,uv);
% plot surface and variation of Jacobian for surface 1
 np=0;
 for nt=1:ntt
   for ns=1:nss
     np=np+1; fprintf(fid,"%8.5f %8.5f %8.5f \n", xyg(1:3,np));
```

```
  if(nsurf == 1)
    norm= vexp(vuv{1}(:,np),vuv{2}(:,np));
    Jac= (norm(1)^2+norm(2)^2+norm(3)^2)^0.5*Jacuv(np);
    fprintf(fid2,"%8.5f %8.5f %8.5f \n",ss(ns),tt(nt),Jac);
  endif
 end
fprintf(fid,"\n");fprintf(fid2,"\n")
 end
fprintf(fid,"\n"); fprintf(fid,"\n");
end
fclose(fid);fclose(fid2);
endfunction;
```

The function is applied to the display of the trimmed cylinder shown in Figure 21. The mapping method is shown graphically in Figure 24. The Jacobian is plotted in Figure 25. It can be seen that the Jacobian tends to zero as the sharp corner of surface 1 is reached. A zero Jacobian can be avoided by shifting the control point of the top curves (see Figure 26).

8   INFINITE NURBS PATCH

In some cases it may be necessary to describe a surface which extends to infinity. Applications are the simulation of a very long tunnel or of the earth surface, which for all practical purposes can be considered of infinite extent.

To model an infinite surface the following geometry description is proposed (see [1]):

$$\mathbf{x} = \sum_{j=1}^{2} \sum_{i=1}^{I} R_{i,j}^{\infty,q}(\mathbf{u}, \mathbf{v}) \cdot \mathbf{x}_{i,j} \tag{46}$$

where

$$R_{i,j}^{\infty,q}(\mathbf{u}, \mathbf{v}) = R_i^q(\mathbf{u}) \cdot M_j^{\infty}(\mathbf{v}) \tag{47}$$

where $R_i^q(\mathbf{u})$ are the one-dimensional NURBS introduced earlier and M_j^{∞} are special shape functions that tend to infinity as v approaches 1:

$$M_1^{\infty}(\mathbf{v}) = \frac{1 - 2\mathbf{v}}{1 - \mathbf{v}} \tag{48}$$

$$M_2^{\infty}(\mathbf{v}) = \frac{\mathbf{v}}{1 - \mathbf{v}} \tag{49}$$

The functions are plotted in Figure 27.

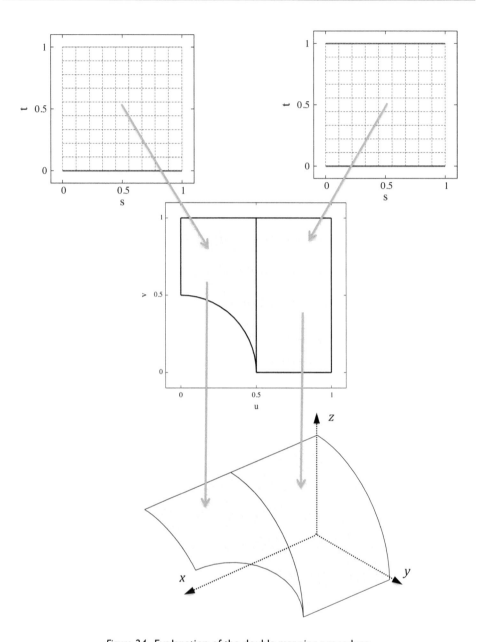

Figure 24 Explanation of the double mapping procedure.

A vector in the tangential direction along u can be computed as

$$\mathbf{V}_1 = \frac{\partial}{\partial \mathbf{u}} \mathbf{x} = \sum_{j=1}^{2} \sum_{i=1}^{I} \frac{\partial}{\partial \mathbf{u}} R_{i,j}^{\infty,q}(\mathbf{u}, \mathbf{v}) \cdot \mathbf{x}_{i,j} \qquad (50)$$

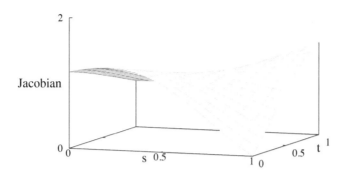

Figure 25 Variation of the Jacobian for trimmed surface I.

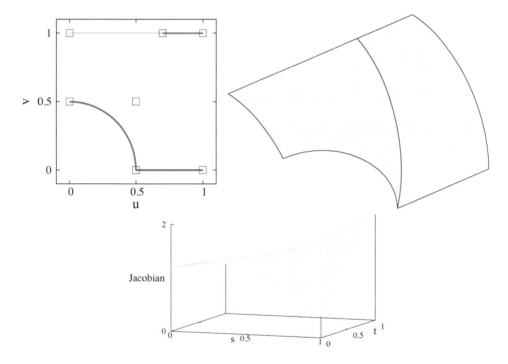

Figure 26 Changed trimming curves, resulting global map and variation of the Jacobian.

and the tangential vector in v-direction

$$V_2 = \frac{\partial}{\partial v}\mathbf{x} = \sum_{j=1}^{2}\sum_{i=1}^{I}\frac{\partial}{\partial v}R_{i,j}^{\infty,q}(u,v)\cdot\mathbf{x}_{i,j} \tag{51}$$

From this the Jacobian can be computed using Equations (29) and (30).

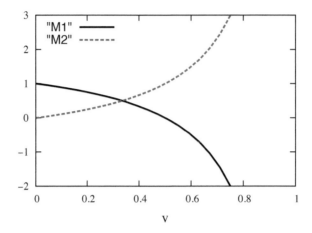

Figure 27 Infinite shape functions.

The derivatives of the functions are given by

$$\frac{\partial}{\partial u} R_{i,j}^{\infty,q}(u,v) = \frac{\partial R_i^q(u)}{\partial u} M_j^\infty(v) \tag{52}$$

$$\frac{\partial}{\partial v} R_{i,j}^{\infty,q}(u,v) = \frac{\partial M_j^\infty(v)}{\partial v} R_i^q(u) \tag{53}$$

where

$$\frac{\partial M_1^\infty}{\partial v} = \frac{-1}{(1-v)^2} \tag{54}$$

$$\frac{\partial M_2^\infty}{\partial v} = \frac{1}{(1-v)^2} \tag{55}$$

It can be seen that the Jacobian approaches infinity as v approaches 1. A function similar to the NURBS toolkit function *nrbdeval* for performing the infinite mapping is shown below.

```
function [xyz,Jac] = nrbdevalInf(nurbs,Coords,tt)
%-----------------------------------------
% Mapping from u,v to x,y,z coordinates
% for infinite NURBS patches
%
% Input:
% nurbs ... structure of NURBS curve on finite edge
% Coords(1:4,n) ...  Coordinates of control points
% tt ... parametric points
```

```
%
% Output:
% xyz   ...  x,y,z -coordinates of param.point
% Jac   ...  derivatives
%-----------------------------------------
ut= tt(1,:);vt= tt(2,:);
ncu= nurbs.number; knotu= nurbs.knots; pu= nurbs.order -1;
%-------------------------
%  Finite functions and derivatives
%-------------------------
B= nrbbasisfun(ut,nurbs);
s = findspan (ncu-1,pu,ut,knotu);
nofun= numbasisfun (s, ut, pu, knotu) + 1 ;
nbfun= columns(nofun);
dB= nrbbasisfunder(ut,nurbs);
%  Note: only non-zero functions are computed
for i=1:length(ut)
 for ncp=1:ncu*2
   Rinf(i,ncp)= 0.; dRinfdu(i,ncp)= 0; dRinfdv(i,ncp)= 0;
 end
end
%    Infinte functions
for i=1:length(vt)
 denom= 1-vt(i); denom2= denom^2;
 M(1)= (1-2*vt(i))/denom; M(2)= vt(i)/denom;
 dM(1)= -1/denom2; dM(2)= 1/denom2;
% Combined functions
 ncp=0;
 for nv=1:2
  for nu=1:ncu
   no= 0;
   for n= 1:nbfun;
     if(nofun(i,n) == nu);no=n ; endif;
   end
   ncp=ncp+1;
   if (no > 0)
     Rinf(i,ncp)= M(nv)*B(i,no);
     dRinfdu(i,ncp)= dM(nv)*B(i,no); dRinfdv(i,ncp)= M(nv)*dB(i,no);
   endif
  end
 end
end
%    Compute coordinates and derivative at parametric point
for i=1:length(ut)
  x(1:3)= 0; dxdu(1:3)= 0; dxdv(1:3)= 0; ncp=0;
  for nv=1:2
    for nu=1:ncu
    ncp= ncp+1;
    for nd=1:3
     x(nd)= x(nd) + Rinf(i,ncp)*Coords(nd,ncp);
     dxdu(nd)= dxdu(nd) + dRinfdu(i,ncp)*Coords(nd,ncp);
```

```
      dxdv(nd)= dxdv(nd) + dRinfdv(i,ncp)*Coords(nd,ncp);
    end
   end
 end
 xyz(:,i)= x; v1(:,i)= dxdu; v2(:,i)= dxdv;
end
Jac{1}= v1; Jac{2}= v2;
endfunction
```

8.1   Example

As an example a quarter cylinder of infinite extent is presented. The input data for *nrbdevalInf* are:

```
nurbs= structure for quarter circle with radius 1
Coords:
0 0 0 1 1 1
1 1 0 1 1 0
0 1 1 0 1 1
```

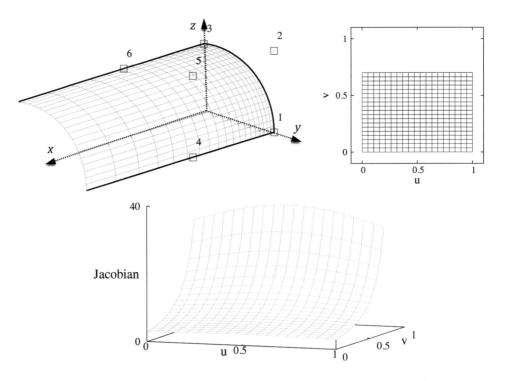

Figure 28 Mapping of infinite NURBS patch from u, v to x, y, z coordinates and variation of the Jacobian.

Figure 28 shows the mapping of a quarter cylinder with infinite extension in the x-coordinate direction and the variation of the Jacobian. It can be seen that it rises sharply in the infinite (v) direction.

9   SUMMARY AND CONCLUSIONS

In this chapter we have discussed one of the most important aspects of numerical simulation, namely the description of the geometry. By several examples we have demonstrated that the definition of geometry with NURBS basis functions is far superior to its approximation with Lagrange or Srendipity functions, which are currently used in software. An important milestone has been reached, namely the efficient and accurate definition of geometry with few parameters. The next stage is concerned with understanding how CAD programs operate and the type of output we can expect from them.

BIBLIOGRAPHY

[1]   G. Beer. Mapped infinite patches for the NURBS based boundary element analysis in geomechanics. *Computers and Geotechnics*, 66:66–74, 2015.

[2]   G. Beer, B. Marussig, and J. Zechner. A simple approach to the numerical simulation with trimmed CAD surfaces. *Computer Methods in Applied Mechanics and Engineering*, 285:776–790, 2015.

[3]   H.-J. Kim, Y.-D. Seo, and S.-K. Youn. Isogeometric analysis for trimmed CAD surfaces. *Computer Methods in Applied Mechanics and Engineering*, 198(37–40):2982–2995, 2009.

[4]   Hyun-Jung Kim, Yu-Deok Seo, and Sung-Kie Youn. Isogeometric analysis with trimming technique for problems of arbitrary complex topology. *Computer Methods in Applied Mechanics and Engineering*, 199(45–48):2796–2812, 2010.

[5]   R. Schmidt, R. Wuechner, and K. Bletzinger. Isogeometric analysis of trimmed NURBS geometries. *Computer Methods in Applied Mechanics and Engineering*, 241–244:93–111, 2012.

[6]   Matt Sederberg. *T-splines3 users manual*. T-splines inc., 34 E 1700 S Suite A143 Provo, UT 84606, www.tsplines.com edition, 2011.

Chapter 4

Stage 3: Computer Aided Design

Imagination is more important than knowledge

A. Einstein

The aim of this book is to facilitate the development of software that uses geometrical data directly from CAD programs, without the need to generate a mesh. Since the main aim of CAD programs is to display geometry, the data produced by them are not immediately suitable for numerical simulation. In the next part of our journey we look inside CAD programs and investigate ways of using the data for simulation.

I INTRODUCTION

The beginning of CAD can be traced to the year 1957, when Dr. Patrick J. Hanratty developed PRONTO, the first commercial numerical-control programming system. In 1960 Ivan Sutherland at MIT's Lincoln Laboratory created SKETCH-PAD, which demonstrated the basic principles and feasibility of computer technical drawing.

The first CAD systems served as mere replacements of drawing boards. The design engineer still worked in 2D to create technical drawing consisting of 2D wireframe primitives (line, arc, B spline ...). 3D wireframe features were developed in the beginning of the sixties, and in 1969 the first commercially available solid modeler program was released, further enhancing the 3D capabilities of CAD systems. Mathematical representation of freeform surfaces with NURBS, appeared in 1989 when CADs based on parametric engines were introduced, i.e. a model is defined by parameters. This means that a change of dimension values in one place also changes other dimensions preserving relation of all elements in the design.

CAD applications now offer advanced rendering and animation capabilities, so engineers can better visualize their product designs. 3D immersive technology allows clients to view a building or car before it is made. 4D BIM is a type of virtual construction engineering simulation incorporating time or schedule related information for project management. CAD has become an especially important technology within the scope of computer-aided technologies, with benefits such as lower product

Figure 1 Example of computer aided design.

development costs and a greatly shortened design cycle [2]. Figure 1 shows an example of a car designed with program Rhino[1].

The main purpose of CAD programs is therefore the manipulation and display of geometry and it is not concerned with numerical simulation. However, in most cases numerical simulation is an important aspect of design in order for example to test the strength and stability of a structure or to test the aerodynamics of a plane.

In the last decade powerful mesh generators were developed that read geometry data from CAD programs and generate meshes, that approximate this geometry with various degrees of accuracy. It is obvious, that this is not an efficient process. If an accurate geometrical description is available, why does one need to approximate it? As has been demonstrated in the introduction, the process of generating a mesh involves additional work and is also a major source of errors. It is estimated that more than half of the total analysis time may be spent on mesh generation.

The main reason for this state of affairs is that CAD and simulation has developed completely independently and there was no dialogue between the two communities. Because of the initiative of T. Hughes this changed in 2005 with the publication of a paper that for the first time explored how the methods used by the CAD community could also be used for simulation [1]. It was soon discovered that the functions used by the CAD community (NURBS) were not only much better suited than the Lagrange polynomials to describe the geometry of a problem but were also better for the approximation of the unknown.

Since the aim of this book is to use CAD data directly for simulation without the intermediate step of mesh generation, it is important to understand the way CAD programs work and what kind of data are produced. CAD programs communicate with the outside world with various formats, some of them binary and some in text (ASCII) form. Here we focus on a standardized text format, namely the Initial Graphics Exchange Specification (IGES). IGES was an initiative of the United States Air Force (USAF) Integrated Computer Aided Manufacturing (ICAM) project (1976–1984). ICAM sought to develop procedures, processes and software (CAD/CAM) that would integrate all operations in aerospace manufacturing and thus greatly reduce costs. The last document on the IGES standard, produced in 2006 [3], has more than 600 pages.

---

[1]Program Rhinoceros NURBS modeling for Windows, by Robert McNeal and associates.

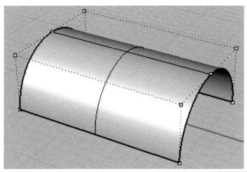

```
                                                                      S    1
1H,,1H;,,                                                             G    1
28HZ:\Documents\IsoBEM\Qcyl.igs,                                     G    2
26HRhinoceros ( May 12 2010 ),31HTrout Lake IGES 012 May 12 2010,    G    3
32,38,6,308,15,                                                      G    4   General
,                                                                    G    5   information
1.0D0,2,2HMM,1,0.254D0,13H140317.235051,                             G    6
0.001D0,                                                             G    7
15D0,                                                                G    8
,                                                                    G    9
,                                                                    G   10
10,0,13H140317.235051;                                               G   11
    314     1      0      0      0      0      0    0000002200D    1
    314     0      1      1      0      0      0  COLOR      0D    2   Colour
    406     2      0      0      1      0      0    0000003000D    3   definition etc.
    406     0     -1      1      3      0    0LEVELDEF     0D    4
    128     3      0      0      1      0      0    0000000000D    5   NURBS surface
    128     0     -1     11      0      0      0  TrimSrf    0D    6
    126    14      0      0      1      0      0    0000000000D    7   NURBS curve
    126     0     -1      7      2      0    03d_BsCrv     0D    8
314,0.0,0.0,0.0,20HRGB(    0,    0,    0 );                  0000001P    1
406,2,1,7HDefault;                                          0000003P    2
128,4,1,2,1,0,0,0,0,0,0.0D0,0.0D0,0.0D0,7.853981633974483D0,  0000005P    3
7.853981633974483D0,15.70796326794897D0,15.70796326794897D0,  0000005P    4   Data
15.70796326794897D0,0.0D0,15.0D0,15.0D0,1.0D0,               0000005P    5   for
0.7071067811865476D0,1.0D0,0.7071067811865476D0,1.0D0,1.0D0,  0000005P    6   NURBS
0.7071067811865476D0,1.0D0,0.7071067811865476D0,1.0D0,0.0D0,  0000005P    7   surface
5.0D0,0.0D0,0.0D0,5.0D0,4.999999999999999D0,0.0D0,           0000005P    8
3.061616997868383D-16,5.0D0,0.0D0,-4.99999999999999D0,5.0D0,  0000005P    9
0.0D0,-5.0D0,6.123233995736766D-16,5.0D0,0.0D0,15.0D0,       0000005P   10
5.0D0,4.999999999999999D0,15.0D0,3.061616997868383D-16,5.0D0, 0000005P   11
15.0D0,-4.999999999999999D0,5.0D0,15.0D0,-5.0D0,             0000005P   12
6.123233995736766D-16,0.0D0,15.70796326794897D0,0.0D0,15.0D0; 0000005P   13
126,4,2,1,0,0,0,0.0D0,0.0D0,0.0D0,7.853981633974483D0,       0000007P   14
7.853981633974483D0,15.70796326794897D0,15.70796326794897D0,  0000007P   15   Data
15.70796326794897D0,1.0D0,0.7071067811865476D0,1.0D0,        0000007P   16   for
0.7071067811865476D0,1.0D0,0.0D0,5.0D0,0.0D0,0.0D0,5.0D0,    0000007P   17   NURBS
4.999999999999999D0,0.0D0,3.061616997868383D-16,5.0D0,0.0D0,  0000007P   18   curve
-4.999999999999999D0,5.0D0,0.0D0,-5.0D0,6.123233995736766D-16, 0000007P  19
0.0D0,15.70796326794897D0,1.0D0,0.0D0,0.0D0,0.0D0;           0000007P   20
S0000001G0000011D0000008P0000020                                      T    1
```

Figure 2   Simple example for explaining the IGES file format, perspective view display of program Rhino with control points and control polygon as well as the IGES file generated.

It should be noted here that the conversion of CAD data into analysis suitable form is a complex task. Here we make an attempt to convert some relatively simple geometries into analysis suitable form, but a lot of work still has to be done in order to allow a seamless integration of CAD and simulation. To the authors's best knowledge so far the main efforts in integrating the two systems has been from the simulation community. In order to achieve a seamless integration avoiding the need of mesh generation perhaps an effort by the CAD community is also required.

2   IGES DATA STRUCTURE

To explain the way CAD programs generate geometry data, let us start with a simple example. A cylindrical surface can be quickly generated using the arc and lofting options of program Rhino and is internally described by one NURBS surface. Figure 2 shows the surface generated in perspective view and the data file generated in IGES format. The data are in ASCII format with a fixed column width of 80 columns with the following features:

- The first part of the file contains general information, which is of no interest to simulation, such as the program that generated the data, the units and the tolerances used etc.
- The second part (starting with 128 in the example) contains data of interest to simulation.

For the second part the following information is supplied in 80 character lines: Information about the object is contained in columns 1 to 64. The first number in the first line relating to an object refers to the key[2] and is followed by data describing the object (details will follow). Data are separated by a comma and a semicolon signifies the end of the data. Columns 66 to 73 are reserved for a pointer that points to the object. This pointer is repeated until the end of data for the object has been reached. Columns 74 to 80 are reserved for a line number.

Our first task is to develop an Octave function that allows us to parse an IGES file looking for certain objects defined by either keys or pointers. Depending on the input, the function Parse has the following functions:

- If a key is specified it finds the pointers to this key item and outputs the pointer numbers
- If a pointer is specified it outputs all information relating to it

The information is stored in an array "Info". The indicator "found" specifies if information has been found for a specified pointer and "Pntrs" provides pointer numbers to all occurrences of a key item. Function *Parseskip* is provided that advances to the first useful line in the IGES file.

```
function Data=Parseskip(fid)
%------------------
%  skips IGES file to first line with relevant data
%------------------
for n=1:10000
  Data= fgetl(fid); if(Data(1) == " ") continue endif
  String= sscanf(Data(1:4),"%s");
  if(String == "110," ) return endif; if(String == "120," ) return endif
  if(String == "126," ) return endif; if(String == "128," ) return endif
```

---

[2]A key refers to an entity type as described later.

```
  if(String == "141," ) return endif
  printf("%s \n","No useful data found")
end
endfunction
```

```
function [Info,found,Pntrs]= Parse(fid,key,Pointer)
%------------------
%  parses IGES file
%  Input: fid      ... ID of IGES file
%         key      ... IGES key (if zero, items for Pointer are read)
%         Pointer  ... Pointer to key item (if zero, all key items are read)
%  Output: Info    ... Data
%          found   ... indicator =1 if specified pointer found
%          Pntrs   ... pointers to found items for specified key
%----------------
i=0; np=0;found = 0; Info(1)=0. ; Pntrs(1)=0 ;
%  skip to first line with relevant Data
Line=Parseskip(fid);
%  look for specified key, reading each line in 3 parts
for n=1:10000
 if(n > 1)
  [Data,nv]= fscanf(fid,"%s",1); if(nv == 0) return endif
  Ptr= fscanf(fid,"%s",1); Lnr= fscanf(fid,"%s",1);
 else
  Data= Line(1:64); Ptr= Line(66:73); Lnr= Line(74:80);
 endif
 [str]=strtok(Data,","); Key= sscanf(str,"%f"); % extract first number in line
 if(key > 0 && Key != key) continue endif
  if(length(Ptr) != 8) return endif  % last line
  [str]=strtok(Ptr,"P");          % Pointer string
  Pntr=sscanf(str,"%f");          % extract pointer number
  np=np+1; Pntrs(np)= Pntr;
  if(Pntr != Pointer) continue endif
  found= 1 ;  break                % data for Pointer found
end
if(found == 0) return endif
%  Extract NURBS information
for n=1:10000
 Rem= sscanf(Data,"%s");
 nch= length(Data);
 for j=1:nch
   [str,Rem]=strtok(Rem,",;");i=i+1;indx= strfind(str,"D");
   if(indx == [])
     Info(i)= sscanf(str,"%f");
   else                              %  read double precision number
     [number,exp]= strtok(str,"D");nch= length(exp);
     ex= sscanf(exp(2:nch),"%f"); Info(i)= sscanf(number,"%f");
     if(ex != 0) Info(i)= Info(i)^ex; endif
   endif
   if(Rem == ";") return endif  %  end of data
   if(Rem == ",") break endif   %  next line
 end
 Data= fscanf(fid,"%s",1); Ptr= fscanf(fid,"%s",1); Lnr= fscanf(fid,"%s",1);
end
endfunction
```

3   HOW CAD PROGRAMS DESCRIBE GEOMETRY – ENTITY TYPES

CAD programs describe curves and surfaces in a number of ways. Solids are described by their bounding surfaces. For the description of surfaces *Entity types* are used and these were referred to as key items previously. There are 147 entity types listed in the IGES manual. Only few of them actually deal with the description of curves and surfaces by NURBS. Here we describe some entities that will be used later in the book. Info relates to the array returned by function Parse.

Entity types

- Type 110: Line entity
- Type 120: Surface of revolution entity
- Type 126: Rational B-spline entity
- Type 128: Rational B-spline surface entity
- Type 141: Boundary entity

3.1   Line entity (110)

The line entity relates to the definition of a line by specifying the start and end point coordinates. This will be used later to define the axis of revolution for surfaces of revolution.

Six parameters are provided:

- Info:

    Info(1): key = 110
    Info(2): x-coordinate of start of line
    Info(3): y-coordinate of start of line
    Info(4): z-coordinate of start of line
    Info(5): x-coordinate of end of line
    Info(6): y-coordinate of end of line
    Info(7): z-coordinate of end of line

3.2   Surface of revolution entity (120)

This entity defines a surface of revolution as explained in the previous stage. The definitions of the start and end angle are the same as explained there.

Four parameters are provided:

- Info:

    Info(1) = key = 120
    Info(2) = Pointer to the line entity specifying the axis of revolution
    Info(3) = Pointer to the generatrix entity
    Info(4) = Start angle in radians (α_s)
    Info(5) = End angle in radians (α_e)

3.3   Rational B-spline entity (126)

This provides information about a NURBS curve.

- Info:

   Info(1): key = 126,
   Info(2): I (ncu = I + 1: Number of control points in u-direction)
   Info(3): pu, order in u-direction
   Info(4:7): Properties of curve
   Info(...): knotu, Knot vector in u-direction
   Info(...): Weights (w(0) .... w(I))
   Info(...): Coordinates: (x(0), y(0), z(0), ..., x(I), y(I), z(I))

Notes: Properties of curve have no immediate relevance here. Note that control points are numbered starting from zero.

3.4   Rational B-spline surface entity (128)

This provides information about a NURBS surface.

- Info:

   Info(1): Key = 128,
   Info(2): I (ncu= I + 1: Number of control points in u-direction)
   Info(3): J (ncv = J + 1: Number of control points in v-direction)
   Info(4): pu, order in u-direction
   Info(5): pv, order in v-direction
   Info(6:10): Properties of surface
   Info(...): knotu, Knot vector in u-direction
   Info(...): knotv, Knot vector in v-direction
   Info(...): Weights (w(0,0),w(1,0) .... w(I,J))
   Info(...): Coordinates: (x(0,0), y(0,0), z(0,0), ..., x(I,J), y(I,J), z(I,J))
   Info(...): *ustart, uend, vstart, vend*

Notes: Properties of surface have no immediate relevance here. *ustart, uend* and *vstart, vend* refer to the first and last parameter values in the knotu, knotv vectors. CAD programs are quite liberal with the entries in the knot vectors. Sometimes they start with negative numbers and entries do not range from 0 to 1. This information provides the values of u, v at the start and end of the parameter space and is required for trimmed surfaces.

3.5   Boundary entity (141)

This provides information for trimmed NURBS surfaces. The information given consists of pointers to the (trimming) curves defining the trimmed surface. They are provided in the global (x, y, z) space (specified as model space in the CAD community) and the local parameter space of the NURBS surface to be trimmed. For the trimming we use the method explained in the last stage and we only need the information in parameter space (therefore only the information for the parameter space is shown). The number of trimming curves and the sequence in which curves are specified in the IGES file is quite arbitrary and this makes the extraction of information required for the trimming process more involved. More about this later.

- Info

  Info(1): Key = 141
  Info(2): Type of representation provided (model/parameter or both)
  Info(3): Preferred representation (not sure what this for)
  Info(4): Pointer to surface to be trimmed
  Info(5): ncurvs = Number of trimming curves
  i = 9; skip information provided for global coordinate space
  for n = 1:ncurvs
      i = i + 4
      Info(i) = crv(n) ... Pointers to trimming curves in parameter space
  end

4 NURBS SURFACES

Here we show how data can be read from the IGES file and converted into a form usable for our purposes. We start with a simple NURBS surface and introduce a function Get_infos that reads information of a surface whose location is specified in the IGES file by Pointer.

```
function [knotu,knotv,coefs,ncu,ncv,limit,found] = Get_infos(file,Pointer)
%-----------------------------------
%  Extracts data for NURBS surface
%
%  Input:
%  file  ... name of IGES file
%  Pointer  .. pointer to surface
%
%  Output:
%  knotu,knotv  ...  knot vectors
%  coefs  ...  coefficients (control point coords and weights)
%  ncu,ncv  ...  number of control points
%  limit  ...  limits of parameter space
%  found  ...  indicator if found
%-----------------------------------
fid=fopen(file,"r");
[Info,found] = Parse(fid,0,Pointer); fclose(fid);
if(found == 0) printf("%s %d","no info for pointer", Pointer), return; endif
Type= Info(1);  if(Type != 128) printf("%d %s", Pointer," not a surface");
found= 0; return; endif
i=2; ncu= Info(i)+1;i=i+1; ncv= Info(i)+1;
i=i+1; pu= Info(i); i=i+1; pv= Info(i); i= 10;
```

```
for n=1:ncu+pu+1; i=i+1; knotu(n)= Info(i); end
for n=1:ncv+pv+1; i=i+1; knotv(n)= Info(i); end
% extract weigths
for nv=1:ncv
 for nu=1:ncu
  i=i+1; coefs(4,nu,nv)= Info(i);
 end
end
% extract coefficients
for nv=1:ncv
 for nu=1:ncu
  w=coefs(4,nu,nv);
  i=i+1; coefs(1,nu,nv)= Info(i)*w ;
  i=i+1; coefs(2,nu,nv)= Info(i)*w; i=i+1; coefs(3,nu,nv)= Info(i)*w;
 end
end
% read paramter limits (for trimming)
i=i+1; limit(1)= Info(i); i=i+1; limit(2)= Info(i);
i=i+1; limit(3)= Info(i); i=i+1; limit(4)= Info(i);
endfunction;
```

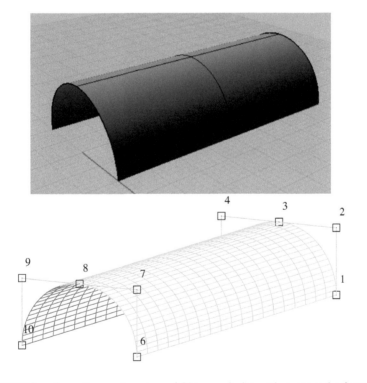

Figure 3 A NURBS surface. Perspective view of Rhino and plot with extracted information using function PlotNURBSurf.

Next we show function PlotNURBSurf that can be used to plot a surface specified by Pointer.

```
function PlotNURBSurf(file,Pointer)
%--------------------------------------
%    Creates data for plotting a NURBS surface
%    with information read from IGES file
%    Input:
%    file ... IGES file name
%    Pointer ... pointer to surface to be plotted
%--------------------------------------
[knotu,knotv,coefs,ncu,ncv,limit,found] = Get_infos(file,Pointer);
if (found == 0)
 printf("%s %d","Item not found for Pointer",Pointer)
 return
endif
PlotSurface(coefs,knotu,knotv)
```

Figure 3 shows a screenshot of the perspective view in Rhino of the surface and a re-plot using the function PlotNURBSurf showing control point location and their numbering.

5 TRIMMED NURBS SURFACES

The conversion of information about simple geometries that can be described by NURBS surfaces, to analysis suitable form is fairly straight forward. However, more complex geometries often involve a union of 2 surfaces. Examples of a union between two cylindrical surfaces can be seen in Figure 4.

There are two options that can be used in Rhino:

- Sharp corners exist at the junction
- There is a smooth transition (fillet) between the surfaces

Rhino generates the graphical display by trimming surfaces that are connected. Therefore, the information provided in the IGES file consists of NURBS data for untrimmed surfaces and for trimming curves. In the first option there are 2 trimmed NURBS surfaces, in the second option there are 2 trimmed and 1 untrimmed (fillet) surface.

The data for the trimming curves are stored under the key item 141. Next we develop functions that read the IGES data and provide the information in the format required by the trimming functions introduced previously. The function Get_Trim is provided to get information about the trimming curves for the surface to be trimmed (defined by Pointer). "Cpointer" contains pointers to trimming curves and "ncurvs" is the number of trimming curves provided.

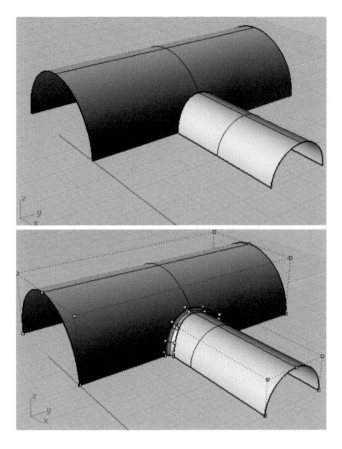

Figure 4 Example of a union between two surfaces with and without fillet.

```
function [Cpointer,ncurvs]=Get_Trim(file,Pointer)
%--------------------------------------
% Gets trimming information for a NURBS surface
%
% Input:
% file    ...   IGES file name
% Pointer ...   Pointer to NURBS surface
%
% Output:
% Cpointer ...   pointers to trimming curves
% ncurvs ... number of trimming curves
%--------------------------------------
fid=fopen(file,"r");[Info,found,Pntrs]=Parse(fid,141,0);fclose(fid);
nptr= length(Pntrs); ncurvs= 0; nkt=0; nct=0;
```

```
Cpointer(1)=0;
for npt=1:nptr
 Pntr= Pntrs(npt);
 fid=fopen(file,"r");[Info,found]= Parse(fid,0,Pntr);fclose(fid);
 surface= Info(4); if(surface != Pointer) continue endif
 ncurvs= Info(5);
 i=9;
 for n=1:ncurvs
  Cpointer(n)= Info(i);i=i+4;
 end
 break
end
endfunction;
```

The function Read_trim reads the data for a trimming curve (control points, weights[3] and knot vector) and adjusts the control point coordinates so they fit within the parameter space of the surface to be trimmed. *Pointer* points to the trimming curve to be read and *limit* is an array that contains the limits of the parameter space of the trimmed NURBS surface.

```
function [knotu,coefs,ncu]=Read_trim(file,Pointer,limit)
%------------------------------------
%   Reads trimming curve data and adjust them
%   to fit the parameter space of the trimmed surface
%
%   Input:
%   file   ...  name of IGES file
%   Pointer ...  pointer to trimming curve
%   limit ... limits of surface parameter space
%
%
%   Output:
%   knotu  ...  knot vector
%   coefs  ...  coefficients
%   ncu ... number of control points
%------------------------------------
spanu= limit(2) - limit(1); spanv= limit(4) - limit(3);
fid=fopen(file,"r"); [Info,found] = Parse(fid,0,Pointer);fclose(fid);
i=2; ncu= Info(i)+1; i=i+1; pu= Info(i); i= 7;
for n=1:ncu+pu+1; i=i+1; knotu(n)= Info(i);end
% read weigths
```

---

[3]Rhino and most CAD programs use B-splines for the trimming curves so the weights are always 1.

```
for nu=1:ncu;   i=i+1; coefs(4,nu)= Info(i); end
% read coefficients
for nu=1:ncu
    i=i+1; coefs(1,nu)= Info(i);
    i=i+1; coefs(2,nu)= Info(i); i=i+1; coefs(3,nu)= Info(i);
end
% adjust coefficients to fit in trimmed surface parameter space
for nu=1:ncu
    if(spanu != 0 && spanv != 0)
    coefs(1,nu)= (coefs(1,nu) - limit(1))/spanu;
    coefs(2,nu)= (coefs(2,nu) - limit(3))/spanv;
    endif
end
endfunction;
```

Using the function *PlotSurface* introduced in the last stage, we can plot the untrimmed surfaces in Figure 5, for the case with the fillet. One has to be careful, because the number of control points for the trimming curves can be unreasonably high. For example without changing the default accuracy the number of control points for trimming curve 1 would have been 90 (!!). In program Rhino the tolerance can be reduced in *preferences* under *units*. With this reduced tolerance a reasonable number of control points can be achieved.

Next we introduce a function *Plot_trcurves* that plots the trimming curves for a surface. The function takes *nsurf* as input and this relates to the sequence in which the surfaces have been found in the IGES file. It generates 4 files for plotting: One describing the shape of the curves, one the location of control points, one specifying the tangent to the curve and the last one the sequence number of the curve. The last two items are needed so that the trimming information can be supplied in suitable form to the trimming functions developed in the last stage.

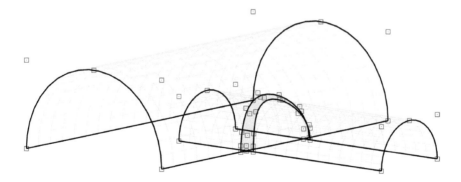

Figure 5 Display of untrimmed surfaces for the case with a fillet.

```
function Plot_trcurves(file,nsurf)
%-------------------------------------
%    Creates data for plotting trimming curves
%    for surface nsurf
%
%    Input:
%    file ... name of IGES file
%    nsurf  ...  sequence of surface found
%
%         .
%    Output to files:
%    "Geometry"  ... contains geometry data
%    "Control" ... contains control point data
%    "Tangent" ... direction of tangent vector
%    "Number" ... info for curve number
%    If surface is not trimmed files are empty
%-------------------------------------
lut=20; lum= 10; ut= linspace(0,1,lut);
fid=fopen(file,"r"); [Info,found,Pntrs]= Parse(fid,128,0);fclose(fid);
Pointer= Pntrs(nsurf)
[knotu,knotv,coefs,ncu,ncv,limit,found] = Get_infos(file,Pointer);
[Cpointer,ncurvs]=Get_Trim(file,Pointer);
if(ncurvs == 0)
 printf("%s \n","Surface not trimmed")
 return
endif
fid=fopen("Geometry","w");fid1=fopen("Control","w");
fid2=fopen("Tangent","w");fid3=fopen("Number","w");
nct= 0;
for ncrv=1:ncurvs
  clear knotu; clear coefs;
  [knotu,coefs,ncu]=Read_trim(file,Cpointer(ncrv),limit);
  for nc=1:ncu
    nct= nct+1;
    fprintf(fid1,"%8.5f %8.5f %d \n", coefs(1:2,nc),nct);
  end
  nurbs= nrbmak(coefs,knotu);
  xyg= nrbeval(nurbs,ut);
  for nu=1:lut
    fprintf(fid,"%8.5f %8.5f %d \n", xyg(1:2,nu),ncrv);
  end
  fprintf(fid,"\n"); fprintf(fid,"\n");
  dnurbs= nrbderiv(nurbs);
  [xym,vuv]= nrbdeval(nurbs,dnurbs,ut(lum));
  tan(1)= vuv(1,1)*0.25; tan(2)= vuv(2,1)*0.25;
  fprintf(fid2,"%8.5f %8.5f %8.5f %8.5f \n", xym(1:2,1),tan(1:2));
  fprintf(fid3,"%8.5f %8.5f %d \n", xym(1:2,1),ncrv);
end
printf("%s %d %s \n","Info generated for",ncurvs," trimming curves")
fclose(fid);fclose(fid1);fclose(fid2);fclose(fid3);
endfunction;
```

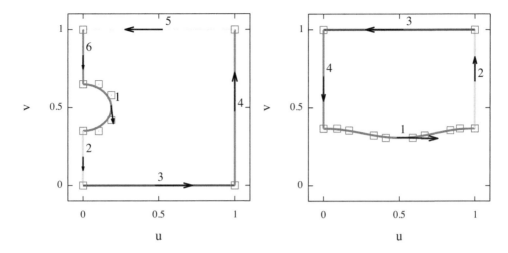

Figure 6 Trimming curves for the cylindrical surfaces with control points and tangent vectors. Numbers indicate sequence in which the curves are output.

The trimming curves associated with the two cylindrical surfaces are plotted using this function in Figure 6. As can be seen, the way we receive trimming information is not immediately suitable for the trimming functions developed in stage 2. Firstly the control points are numbered in anti-clockwise sense and therefore the directions of the tangent vectors are not always in the direction of the local coordinates. The sequence in which the curves are received is anti-clockwise, but as we can see in Figure 6 the starting point is quite arbitrary. This is a quite challenging aspect to solve, if we want a seamless integration of CAD and simulation. Here we attempt an automatic conversion of the data to a suitable form for the trimming software developed previously. The first task is to select the appropriate curves that are opposite to each other and to determine the direction of the linear interpolation.

The strategy is as follows:

- First we read the trimming curve information. It is important to check if the control point sequence is in the direction of u or v coordinates. If it is not the order has to be reversed (see function *Reverse* later on).
- Next we select two opposing curves for the interpolation. The criterion we use is that two linear curves that are parallel in either u or v directions are eliminated and the remaining curves are selected.
- Next we have to determine the direction of the linear interpolation. For example, if the selected curves span across the u-direction, then the interpolation direction is in the v-direction.
- Finally we have to determine which of the curves is on *top* and which one at the *bottom*.

For the case that more than two trimming curves are selected, the region has to be sub-divided into sub-regions as explained in stage 2. The number of subregions is specified by the input variable *ntrims*.

```
function [Ncurv,loc]= Trimcurvs(file,ncurves,Cpointer,limit,ntrims)
%---------------------------------------
%     Selects the trimming curves used for the mapping
%     Splits trimming curves if there is more than 1 subregion
%
%     Input:
%     file ... IGES file name
%     ncurves ... Number of trimming curves
%     Cpointer  ... array of pointers to trimming curves
%     limit ...  starting and ending value of parameter space
%     trims  ...   number of subregions
%
%     Output:
%     Ncurv  ...  Array with pointers to selected curves
%     loc ... direction of interpolation
%-----------------------------------
for ncurv=1:ncurves
 Pointer= Cpointer(ncurv);
 [knotu,coefs,ncu]=Read_trim(file,Pointer,limit);
 [knotu,coefs,si]=Reverse(knotu,coefs,ncu);  % reverse sense if necessary
%-------------------------------
%    find curves along u,v=0,1,
%    code=-1 along v=0, code=+1 along v=1
%    code=-2 along u=0, code=+1 along u=1
%-----------------------------
 r= Dist2(coefs(:,1),coefs(:,2));
 loco(ncurv)= 0; rev(ncurv)=si;
 if(ncu == 2 && r == 1)  % only consider lines that span 1
  if(coefs(1,1) == 0 && coefs(1,2) == 0) loco(ncurv)=-1; endif
  if(coefs(1,1) == 1 && coefs(1,2) == 1) loco(ncurv)=1; endif
  if(coefs(2,1) == 0 && coefs(2,2) == 0) loco(ncurv)=-2; endif
  if(coefs(2,1) == 1 && coefs(2,2) == 1) loco(ncurv)=2; endif
 endif
end
% how many parallel curves in u,v direction ?
ncrvu=0;ncrvv=0;
for ncurv=1:ncurves
 if(abs(loco(ncurv)) == 1) ncrvv=ncrvv+1; endif
 if(abs(loco(ncurv)) == 2) ncrvu=ncrvu+1; endif
end
% if there is only single occurrance > select as one of the curves
for ncurv=1:ncurves
 if(ncrvv == 1 && abs(loco(ncurv)) == 1) ncurve1=ncurv; loc=1; endif
 if(ncrvu == 1 && abs(loco(ncurv)) == 2) ncurve1=ncurv; loc=2; endif
end
% loc=1: interpolation in u-direction; loc=2: in v-direction
ncurve2= ncurve1+2;  % second curve
if(ncurve2 > ncurves) ncurve2= ncurve2 - ncurves; endif
% 1. and 2. curves
if(loco(ncurv) < 0) nl=2; nr=1; else nl=1; nr=2;endif
if(ntrims == 1)
%-------------
% 1 subregion
%-----------
  Ncurv(nl,1)= Cpointer(ncurve1); Ncurv(nr,1)= Cpointer(ncurve2);
```

```
else
%-------------
%  3 subregions
%-----------
 for nsu=1:ntrims
   Ncurv(nr,nsu)= Cpointer(ncurve1); Ncurv(nl,nsu)= Cpointer(ncurve2);
   if(nsu == ntrims) break endif
   ncurve2= ncurve2+1;if(ncurve2 > ncurves) ncurve2= ncurve2 - ncurves; endif
 end
 if(rev(ncurve2) == 1)
% reverse order of curves
   temp= Ncurv(nl,1); Ncurv(nl,1)=Ncurv(nl,3); Ncurv(nl,3)=temp;
  endif
endif
endfunction;
```

The function Reverse listed below checks if the sequence of the control points is in the direction of the local axes and reverses it if it is not.

The criterion introduced here, assumes that the sequence of the control points can be deduced from the tangent to the curve. This should work for most cases but is by no means foolproof.

```
function [knotur,coefsr,si]=Reverse(knotu,coefs,ncu)
%-----------------------------------
%     Reverses curve info if necessary
%
%    Input:
%    knotu ...  kot vector
%    coefs ...   coefficient
%    ncu  ...   number of control points
%
%    Output:
%    knotur  ...  updated knot vector
%    coefsr  ...  updated coefficients
%    si  ...  (0= unchanged; 1= reversed
%-----------------------------------
si=0;nurbs= nrbmak(coefs,knotu); dnurbs= nrbderiv(nurbs);
[xym,vuv]= nrbdeval(nurbs,dnurbs,0.5);
tan(1)= vuv(1,1); tan(2)= vuv(2,1);
rev=0;
if(abs(tan(1)) > abs(tan(2)))
   if(tan(1) < 0) rev=1; endif
else
   if(tan(2) < 0) rev=1; endif
endif
if(rev == 0) knotur=knotu;coefsr=coefs; return; endif
si=1;
nu=length(knotu)+1;
```

```
for n=1:length(knotu)
 nu= nu-1;
 knotur(n)= knotu(nu);
end
nu= ncu+1;
for n=1:ncu
 nu=nu-1;
 coefsr(1:4,n)=coefs(1:4,nu);
end
endfunction;
```

We can now develop a function Plot_surfacetrim that creates files containing information for plotting trimmed and untrimmed surfaces.

```
function Plot_surfacetrim(file,notrim,nsurfp)
%------------------------------------
%    Creates data for plotting trimmed surfaces
%    Input:
%    file  ... name of IGES file
%    notrim= 1 ... plot untrimmed
%    nsurfp=   ... surface to be plotted (if 0 plot all surfaces)
%
%    Output to file
%    "Geometry"  ... contains geoemtry data in x,y,z coords
%    "Control"   ... contains control points data
%    "Outline"   ... contains Jacobian data
%------------------------------------
nst=20; st= linspace(0,1,nst);
%  u,v for untrimmed surface
np=0;
for nv=1:nst
 for nu=1:nst
  np=np+1; uv(1,np)= st(nu); uv(2,np)= st(nv);
 end
end
%  read surface information
fid=fopen(file,"r"); [Info,found,Pntrs]= Parse(fid,128,0);fclose(fid);
nsurfaces= length(Pntrs);
%  open plotting files
fid= fopen("Geometry","w");fid1= fopen("Control","w");
fid2= fopen("Outline","w");
nct= 0; surf=0;
for nsurf=1:nsurfaces
 if(nsurfp > 0 && nsurf != nsurfp) continue endif
 Pointer= Pntrs(nsurf);
 clear knotu; clear knotv; clear coefs;
 [knotu,knotv,coefs,ncu,ncv,limit,found] = Get_infos(file,Pointer);
% output control points of surfaces
 for nv=1:ncv
  for nu=1:ncu
   w=coefs(4,ncu,ncv); nct=nct+1;
   fprintf(fid1,"%8.5f %8.5f %8.5f %d \n", coefs(1:3,nu,nv)/w,nct);
  end
```

```
end
ntrims=1;
if(notrim == 1)
  trim= 0;
else
% get pointers to trimming cuves
  [Cpointer,ncurvs]=Get_Trim(file,Pointer);
  if(ncurvs == 0) trim=0; else trim=1; endif
endif
if(trim == 1)
  if(ncurvs > 4) ntrims=3; endif
% select trimming curves for interpolation
  [Ncurv,loc]= Trimcurvs(file,ncurvs,Cpointer,limit,ntrims);
endif
for nsub=1:ntrims
  if(trim == 1)
  [knotu1,coefs1,knotu2,coefs2]= Getcurves(file,ntrims,nsub,Ncurv,limit,loc);
% 1. map from s,t coordinates
  [uv,Jacuv]= Map(knotu1,coefs1,knotu2,coefs2,st,loc);
  endif
% 2. map from u,v to x,y,z coordinates
  nurbs= nrbmak(coefs,{knotu,knotv}); dnurbs= nrbderiv(nurbs);
  [xyg,vuv]= nrbdeval(nurbs,dnurbs,uv);
  np=0;
%-------------------------
%  create plot files
%-------------------------
  for nt=1:nst
   for ns=1:nst
    np=np+1;
    fprintf(fid,"%8.5f %8.5f %8.5f \n", xyg(1:3,np));
   end
   fprintf(fid,"\n");
  end
  fprintf(fid,"\n"); fprintf(fid,"\n");
  Plot_outlinexyz(st,st,xyg,fid2);
 end
 fprintf(fid,"\n"); fprintf(fid2,"\n");
 surf=surf+1;
end
printf("%s %d %s \n","Info generated for",surf," surfaces")
fclose(fid); fclose(fid1); fclose(fid2);
endfunction;
```

Figure 7 shows the mapping of surface number 1, that was divided into 3 subregions. Figure 8 shows a plot of the trimmed surfaces.

6 SUMMARY AND CONCLUSIONS

In this stage we had a close look at how CAD programs work and especially the format of the IGES file, that allows an interchange of information with the simulation program. While treating NURBS surfaces is very simple we discovered that reading and understanding the information for trimmed surfaces is quite challenging.

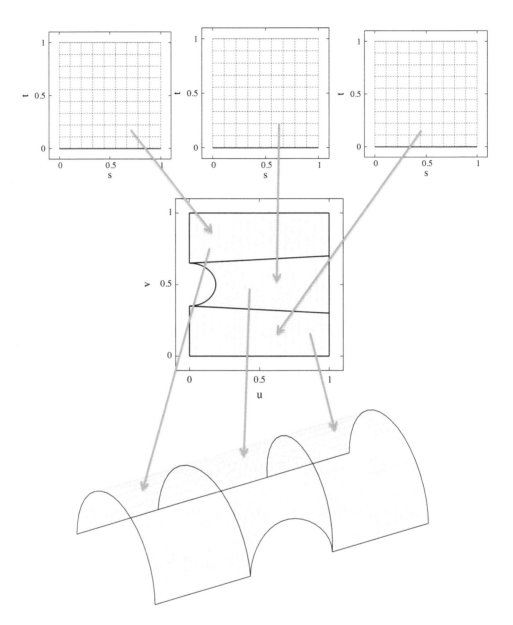

Figure 7 Mapping from local s, t to local u, v and to global x, y, z coordinates for the surface with 3 subregions.

A few words on the modeling of connected surfaces using trimmed NURBS are in order here. Since trimming information is supplied separately for each of the surfaces using their parameter space, there is no guarantee that the surfaces match exactly at the interface. The term used in the community is that the connection is not *watertight*, meaning that there exist small gaps (where presumably water can penetrate). While

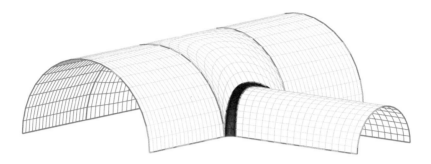

Figure 8 View of geometry with trimmed surfaces.

theoretically this is the case, for practical purposes this is not really relevant. The gaps are so small so that even when zooming in they can not be detected. For the simulation, where much bigger errors are introduced by other aspects such as the approximation of the unknown and the numerical integration this is of no concern.

However, another much more important aspect is that, in some cases, the parameter spaces of the trimming curves may not match at the interface. This will become important when applying this geometry to simulation, as the basis functions at the interface have to match for continuity to be preserved. As we have seen on the example of the NATM tunnel the parameter space is influenced by the knot vector, so so sometimes the problem can be fixed by adjusting the knot values. However, for a seamless integration of CAD and simulation a better solution would have to be found.

We are still far away to find the "holy grail" of a seamless integration of CAD and simulation we hope to have shown a possible path towards attaining this dream.

BIBLIOGRAPHY

[1]   T.J.R. Hughes, J.A. Cottrell, and Y. Bazilevs. Isogeometric analysis: CAD, finite elements, NURBS, exact geometry and mesh refinement. *Computer Methods in Applied Mechanics and Engineering*, 194(39–41):4135–4195, October 2005.

[2]   K. Lalit Narayan. *Computer Aided Design and Manufacturing*. New Delhi: Prentice Hall of India., 2008.

[3]   www.uspro.org/documents/IGES5-3_forDownload.pdf. Iges 5.3, 2006.

Chapter 5

Stage 4: Introduction to numerical simulation

Nature is indifferent towards the difficulties it causes to mathematicians

Fourier

Here we give an introduction to techniques of numerical simulation. The main aim is to become familiar with approximation methods and refinement strategies. We compare Lagrange functions with B-splines and readers should realize at the end of this chapter the advantages of using the latter.

I ONE-DIMENSIONAL SIMULATION

The Bernoulli beam theory is one of the oldest theories in structural mechanics and it is due to Jacob Bernoulli[1]. It allows the simplification of a plane problem to a one-dimensional one. Since things are much easier to explain if there is only one degree of freedom, this is the main reason this problem was chosen to explain the numerical simulation process.

The theory stipulates that when a beam bends, plane sections perpendicular to the beam's axis remain plane and perpendicular to the deformed axis. Consider a beam in plane bending as shown in Figure 1. One can see that when the beam bends upwards fibers below the axis are stretched and above are compressed.

From Figure 1 we compute the change in length, du, of a fiber of length dx at distance z from the axis as

$$du = w'z - (w' - w''dx)z = -w'' \cdot z \cdot dx \qquad (1)$$

where w is the deflection and $w' = \frac{dw}{dx}$, $w'' = \frac{d^2w}{dx^2}$.

The strain in the fiber can be computed by:

$$\epsilon = \frac{du}{dx} = -w'' \cdot z \qquad (2)$$

---

[1]Jacob Bernoulli was born in Basel, Switzerland in 1655. He was appointed professor of mathematics at the University of Basel in 1687, remaining in that position for the rest of his life.

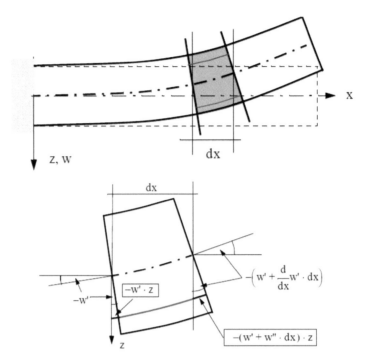

Figure I Beam bending according to Bernoulli.

The resulting stress is:

$$\sigma = E \cdot \epsilon = -E \cdot w'' \cdot z \tag{3}$$

where E is the Modulus of Elasticity.

The internal bending moment can be computed as

$$M = \int_A (\sigma \cdot z)dA = -E \cdot w'' \int_A z^2 \cdot dA = -E \cdot w'' \cdot I \tag{4}$$

where A is the area and the moment of inertia I of the cross-section has been introduced. As the simulation example we choose a beam on elastic foundation shown in Figure 2.

Considering the equilibrium between internal and external forces we obtain the following differential equation (D.E.)[2]:

$$\frac{d^2 M}{dx^2} = q(x) - w \cdot k_w \tag{5}$$

---

[2]For a derivation see for example [5].

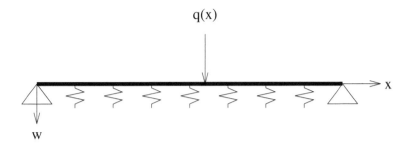

Figure 2 Beam on elastic foundations.

Figure 3 Walter Ritz.

where $q(x)$ is the distributed load and k_w is the stiffness of the elastic foundation. Substitution of Equation 4 gives:

$$\frac{d^2}{dx^2}(EIw'') + w \cdot k_w = q(x) \tag{6}$$

which is a D.E. with respect to x only, i.e. we have reduced the problem dimension by one.

1.1 Ritz method

The D.E. can be solved exactly. However, here we will use this problem to show the process of approximate solution. Therefore, we introduce the concept of a *weak form* solution, where we solve the problem using an approximation for w, $\tilde{w}$ and minimize the error of the solution. This method was first proposed by W. Ritz[3] in 1909 [3]. He proposed to solve the D.E. by assuming an approximation of the solution, using trigonometric functions.

---

[3]Walther Ritz (1878 to 1909) was a Swiss theoretical physicist.

We rewrite Equation (6) as

$$\frac{d^2}{dx^2}(EI\tilde{w}'') + \tilde{w} \cdot k_w - q(x) = R \tag{7}$$

where R is the residual error in satisfying the D.E.

One way of minimizing the error is to multiply the D.E. with a test function δw and to set the integral of the product over the length of the beam L to zero:

$$\int_{x=0}^{L} \left(\frac{d^2}{dx^2}(EI\tilde{w}'') + \tilde{w} \cdot k_w - q(x) \right) \cdot \delta w \cdot dx = 0 \tag{8}$$

This method is also known as *residual method*. Practical applications of the residual method, applicable to structural mechanics, are the method of *virtual work* and the *minimum potential energy*. The first theorem states that for a force system in equilibrium, the work done by real forces and virtual displacements should be zero. The second theorem states that the potential energy of a deformed system in equilibrium should be a minimum. Actually, the second criterion is derived from the first and since the first one is used predominantly in the literature today, this is the one which we will subsequently use.

The beam is subjected to a virtual displacement δw. If the beam is in equilibrium the theorem states that the external and internal work has to sum to zero:

$$\delta W_e + \delta W_i = 0 \tag{9}$$

The work done by a distributed vertical force $q(x)$ is

$$\delta W_e^q = \int_{x=0}^{L} q(x) \cdot \delta w \cdot dx \tag{10}$$

And the work done by the activated spring forces is

$$\delta W_e^k = -\int_{x=0}^{L} \tilde{w} \cdot k_w \cdot \delta w \cdot dx \tag{11}$$

The internal virtual work (*strain energy*) is

$$\delta W_i = -\int_{x=0}^{L} \int_{A} \sigma \cdot \delta\epsilon \cdot dA \cdot dx \tag{12}$$

and after substitution of Eq. 2 and 3 we obtain:

$$\delta W_i = -\int_{x=0}^{L} E \cdot I \cdot \tilde{w}'' \cdot \delta w'' \cdot dx \tag{13}$$

1.2 Approximation

For the approximation it is convenient to use a local coordinate system. We introduce an approximation of w using Lagrange or Serendipity functions as a function of local coordinate ξ:

$$w = \sum_{i=1}^{I} N_i(\xi) \cdot w_i \tag{14}$$

where we drop the $\sim$ and refer to w as being the approximate solution. N_i are basis functions and w_i are *nodal values* of w. The sum is over the number of nodes I. This process is also known as *discretization*. In the case of B-splines[4] the approximation is given as a function of local coordinate u:

$$w = \sum_{i=1}^{I} N_i(u) \cdot w_i \tag{15}$$

where w_i are parameter values and the sum is over the number of control points I. The mapping of the geometry from the local to the global system is given by:

$$x = \sum_{i=1}^{I^g} N_i^g(\xi) \cdot x_i \tag{16}$$

where x_i are the coordinates of the nodes and the superscript g indicates that the basis functions will be independent from the ones used for the approximation of the unknown. This will be referred to as a *geometry independent approximation*.

The Jacobian of this transformation for our one-dimensional beam example is $J = \frac{dx}{d\xi}$. The geometry is exactly defined by linear basis functions.

To proceed, we need the second derivative of w towards x. For Lagrange basis functions this is given by:

$$w'' = \sum_{i=1}^{I} \frac{\partial^2 N_i(\xi)}{\partial x^2} \cdot w_i \tag{17}$$

Applying the chain rule of differentiation we get:

$$\frac{\partial^2 N_i(\xi)}{\partial x^2} = \frac{\partial^2 N_i(\xi)}{\partial \xi^2} \cdot \left(\frac{\partial \xi}{\partial x}\right)^2 + \frac{\partial N_i(\xi)}{\partial \xi} \cdot \frac{\partial^2 \xi}{\partial x^2} \tag{18}$$

with $\frac{\partial \xi}{\partial x} = J^{-1}$.

---

[4]Note that there is no sense in using NURBS in this case as the weights would represent additional unknown values.

The second derivative with B-splines is given by:

$$w'' = \sum_{i=1}^{I} \frac{\partial^2 N_{i,p}(u)}{\partial x^2} \cdot w_i \tag{19}$$

where

$$\frac{\partial^2 N_{i,p}(u)}{\partial x^2} = \frac{\partial^2 N_{i,p}(u)}{\partial u^2} \cdot \left(\frac{\partial u}{\partial x}\right)^2 + \frac{\partial N_{i,p}(u)}{\partial u} \cdot \frac{\partial^2 u}{\partial x^2} \tag{20}$$

with $\frac{\partial u}{\partial x} = J_u^{-1}$ where $J_u = \frac{dx}{du}$ is the Jacobian of the mapping from u to x coordinates. Note that $\frac{\partial^2 \xi}{\partial x^2}$ and $\frac{\partial^2 u}{\partial x^2}$ are zero for a linear mapping of the geometry used for this example.

We explain the method for Lagrange/Serendipity functions first. We subject the beam to a virtual displacement δw_j at node point j (all other values being zero). Substitution of Equation (14) into (10) we get for the virtual work done by the distributed force:

$$\delta W_e^q = \int_{x=0}^{L} q(x)(N_j(\xi) \cdot \delta w_j) \cdot dx = \left(\int_{x=0}^{L} q(x)N_j(\xi) \cdot dx\right) \cdot \delta w_j \tag{21}$$

Similarly for the activated spring forces

$$\delta W_e^k = -\int_{x=0}^{L} \left(k_w \sum_{i=1}^{I} N_i(\xi) \cdot w_i\right) \cdot N_j(\xi) \cdot \delta w_j \cdot dx$$

$$= -k_w \cdot \sum_{i=1}^{I} \int_{x=0}^{L} \left[N_i(\xi) \cdot N_j(\xi) \cdot dx\right] \cdot w_i \cdot \delta w_j \tag{22}$$

For the strain energy we get

$$\delta W_i = -\int_{x=0}^{L} E \cdot I \cdot \left(\sum_{i=1}^{I} \frac{\partial^2 N_i(\xi)}{\partial x^2} \cdot w_i\right) \cdot \frac{\partial^2 N_j(\xi)}{\partial x^2} \cdot \delta w_j \cdot dx \tag{23}$$

or

$$\delta W_i = -E \cdot I \cdot \sum_{i=1}^{I} \left[\left(\int_{x=0}^{L} \frac{\partial^2 N_i(\xi)}{\partial x^2} \cdot \frac{\partial^2 N_j(\xi)}{\partial x^2} \cdot dx\right) \cdot w_i\right] \cdot \delta w_j \tag{24}$$

Adding the internal and external virtual work done by virtual displacements δw_j at all nodes I, yields the following system of equations:

$$\left(\sum_{j=1}^{I} K_{ij} \cdot w_j - F_i\right) \cdot \delta w_i = 0 \quad \text{for } i = 1, \dots, I \tag{25}$$

with the stiffness coefficients K_{ij} and the nodal point forces F_i defined by

$$K_{ij} = \int_{x=0}^{L} \left[E \cdot I \cdot \frac{\partial^2 N_i(\xi)}{\partial x^2} \cdot \frac{\partial^2 N_j(\xi)}{\partial x^2} + k_w \cdot N_i(\xi) \cdot N_j(\xi) \right] \cdot dx \tag{26}$$

$$F_i = \int_{x=0}^{L} q N_j(\xi) dx$$

Since the virtual displacements δw_i are not zero, the value inside the parentheses of Equation (25) has to be zero for the Equation to be satisfied and therefore

$$\sum_{i=1}^{I} K_{ij} \cdot w_j - F_i = 0 \tag{27}$$

In matrix notation we obtain:

$$\mathbf{K} \cdot \mathbf{w} = \mathbf{F} \tag{28}$$

where $\mathbf{K}$ is the stiffness matrix and $\mathbf{F}$ is the force vector, the coefficients of which have been defined in Equation (26) and $\mathbf{w}$ contains nodal displacements. The derivation for B-splines is very similar, except that the values of $\mathbf{w}$ represent parameter values rather than nodal values.

For the evaluation of the integrals we use Gauss numerical integration which states that the integral of a polynomial function can be replaced by a sum of function values at M Gauss points, $f(\xi_m)$, times weights, W_m:

$$\int_{\xi=-1}^{1} f(\xi) \cdot d\xi = \sum_{m=1}^{M} f(\xi_m) \cdot W_m \tag{29}$$

A function *Gauss* that computes the Gauss points and weights for up to 8 points is shown.

```
function [Cor,Wi]= Gauss(ng)
%-------------------------------------------
%    Computes Gauss point coords and weights
%
%    Input:
%    ng   ...   Number of Gauss points
%
%    Output:
%    Cor ...   Gauss point coordinates
%    Wi   ...   Weights
%-------------------------------------------
if(ng == 1);
 Cor(1)= 0.; Wi(1) = 2.0;
```

```
elseif (ng == 2);
 Cor(1)= .577350269; Cor(2)= -Cor(1);  Wi(1) = 1.0; Wi(2) = Wi(1);
elseif(ng == 3);
 Cor(1)= .774596669; Cor(2)= 0.0; Cor(3)= -Cor(1);
 Wi(1) = .555555555; Wi(2) = .888888888; Wi(3) = Wi(1);
elseif(ng == 4);
 Cor(1)= .861136311; Cor(2)= .339981043; Cor(3)= -Cor(2);
 Cor(4)= -Cor(1);
 Wi(1) = .347854845; Wi(2) = .652145154; Wi(3) = Wi(2); Wi(4) = Wi(1);
elseif(ng == 5);
 Cor(1)= .9061798459; Cor(2)= .5384693101; Cor(3)= .0;
 Cor(4)= -Cor(2); Cor(5)= -Cor(1);
 Wi(1) = .236926885; Wi(2) = .478628670; Wi(3) = .568888888;
 Wi(4) = Wi(2); Wi(5) = Wi(1);
elseif(ng == 6);
 Cor(1)= .932469514; Cor(2)= .661209386; Cor(3)= .238619186;
 Cor(4)= -Cor(3);  Cor(5)= -Cor(2); Cor(6)= -Cor(1);
 Wi(1) = .171324492; Wi(2) = .360761573; Wi(3) = .467913934;
 Wi(4) = Wi(3); Wi(5) = Wi(2); Wi(6) = Wi(1);
elseif(ng == 7);
 Cor(1)= .949107912; Cor(2)= .741531185; Cor(3)= .405845151;
 Cor(4)= 0.; Cor(5)= -Cor(3); Cor(6)= -Cor(2); Cor(7)= -Cor(1);
 Wi(1) = .129484966; Wi(2) = .279705391; Wi(3) = .381830050;
 Wi(4) = .417959183; Wi(5) = Wi(3); Wi(6) = Wi(2); Wi(7) = Wi(1);
elseif(ng == 8);
 Cor(1)= .960289856; Cor(2)= .796666477; Cor(3)= .525532409;
 Cor(4)= .183434642; Cor(5)= -Cor(4); Cor(6)= -Cor(3);
 Cor(7)= -Cor(2); Cor(8)= -Cor(1);
 Wi(1) = .101228536; Wi(2) = .222381034; Wi(3) = .313706645;
 Wi(4) = .362683783; Wi(5) = Wi(4); Wi(6) = Wi(3);
 Wi(7) = Wi(2); Wi(8) = Wi(1);
endif;
endfunction
```

The prerequisite of using Gauss integration is that the function is defined within the limits -1 to $+1$. If B-splines are used for the description of w, then an additional mapping is required since u spans from 0 to 1:

$$u(\xi) = \frac{1}{2}(\xi + 1) \tag{30}$$

The Jacobian of this transformation is $\frac{du}{d\xi} = 1/2$. The number of Gauss points that has to be used depends on the order of the function. For example a Gauss order of 3 can evaluate a polynomial of order 5 exactly and 4 Gauss points are sufficient for a polynomial of order 7.

It should be noted here that the integrand involves the Jacobian times second derivatives of the basis function and will in general not be a polynomial. Therefore the Gauss integration will be an approximation, which depends on the number of Gauss points. The integrals, expressed in local coordinate ξ are:

$$K_{ij} = \int_{\xi=-1}^{1} \left[E \cdot I \cdot \frac{\partial^2 N_i(\xi)}{\partial x^2} \cdot \frac{\partial^2 N_j(\xi)}{\partial x^2} + k_w \cdot N_i(\xi) \cdot N_j(\xi) \right] \cdot J \cdot d\xi \tag{31}$$

$$F_j = \int_{\xi=-1}^{1} q(\xi) \cdot N_j(\xi) \cdot J \cdot d\xi$$

Using B-splines for the description of w we have

$$K_{ij} = \int_{\xi=-1}^{1} \left[E \cdot I \cdot \frac{\partial^2 N_i(u)}{\partial x^2} \cdot \frac{\partial^2 N_j(u)}{\partial x^2} + k_w \cdot N_i(u) \cdot N_j(u) \right] \cdot J_u \cdot \frac{du}{d\xi} \cdot d\xi \tag{32}$$

$$F_j = \int_{\xi=-1}^{1} q(u) \cdot N_j(u) \cdot J_u \cdot \frac{du}{d\xi} \cdot d\xi$$

Applying Gauss integration we obtain:

$$K_{ij} = \sum_{m=1}^{M} k_{ij}(\xi_m) \cdot W_m \tag{33}$$

$$F_i = \sum_{m=1}^{M} f_i(\xi_m) \cdot W_m \tag{34}$$

with

$$k_{ij}(\xi) = \left[E \cdot I \cdot \frac{\partial^2 N_i(\xi)}{\partial x^2} \cdot \frac{\partial^2 N_j(\xi)}{\partial x^2} + k_w \cdot N_i(\xi) \cdot N_j(\xi) \right] \cdot J \tag{35}$$

$$f_i(\xi) = q(\xi) \cdot N_j(\xi) \cdot J \tag{36}$$

for Lagrange/Serendipity basis functions and

$$k_{ij}(\xi) = \left[E \cdot I \cdot \frac{\partial^2 N_i(u(\xi))}{\partial x^2} \cdot \frac{\partial^2 N_j(u)}{\partial x^2} + k_w \cdot N_i(u) \cdot N_j(u) \right] \cdot J_u \cdot \frac{du}{d\xi} \tag{37}$$

$$f_i(\xi) = q(u) \cdot N_j(\xi) \cdot J_u \cdot \frac{du}{d\xi} \tag{38}$$

for description by B-splines.

2 STEPS IN THE SIMULATION

Having established the theoretical background we discuss the solution process in more detail. Numerical simulation involves the following steps:

1 Description of the geometry
2 Description of the known values
3 Approximation of the unknown
4 Refinement of the approximation
5 Checking accuracy of the results
6 Display of the results

Steps 4 and 5 should be repeated until the results are of an acceptable accuracy. For demonstrating the simulation process we use the Bernoulli beam example with simple parameters ($L = 1$, $EI = 1/100$) and set the stiffness of the elastic foundation to zero.

2.1 Description of the geometry

For the description of the geometry we can apply a linear basis function and with $x_1 = 0$ and $x_2 = L$:

$$x(\xi) = N_1^g(\xi) \cdot x_1 + N_2^g(\xi) \cdot x_2 \tag{39}$$

The Jacobian is computed by

$$J = \frac{dx}{d\xi} = \frac{dN_1^g(\xi)}{d\xi} \cdot x_1 + \frac{dN_2^g(\xi)}{d\xi} \cdot x_2 = -0.5 \cdot 0 + 0.5 \cdot L = L/2 \tag{40}$$

We can also use a linear B-spline:

$$x(u) = N_{1,1}^g \cdot x_1 + N_{2,1}^g \cdot x_2 \tag{41}$$

The Jacobian is given by:

$$J_u = \frac{dx}{du} = \frac{dN_{1,1}^g(u)}{du} \cdot x_1 + \frac{dN_{2,1}^g(u)}{du} \cdot x_2 = -1 \cdot 0 + 1 \cdot L = L \tag{42}$$

2.2 Description of known values

For checking the convergence we assume a quadratic variation of the loading $q(\xi)$:

$$q(\xi) = \sum_{i=1}^{3} N_i(\xi) \cdot q_i \tag{43}$$

where $N_i(\xi)$ are quadratic Serendipity functions and

$$q_1 = 1; \quad q_2 = 0; \quad q_3 = 1 \tag{44}$$

For this loading the exact solution of w is of order 6.

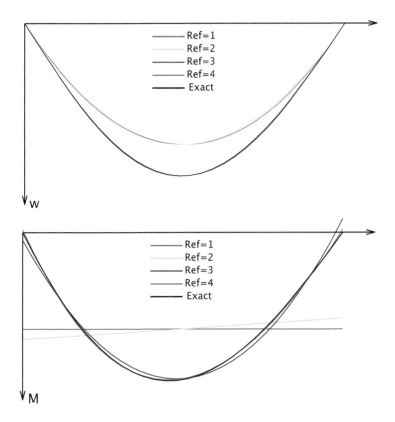

Figure 4 Deflected shapes and bending moment diagrams for the different refinements with increasing order of basis functions (Ref = 1 is order 2 and Ref = 4 is order 5).

2.3 Convergence tests

For checking the accuracy of the solution we define the following error norms:
    Error in the deflection w as

$$\|\epsilon_w\| = \frac{\int_{x=0}^{L} |w - w_{ex}| dx}{\int_{x=0}^{L} |w_{ex}| dx} \tag{45}$$

where w_{ex} is the exact solution.
    Error in the bending moment M as

$$\|\epsilon_M\| = \frac{\int_{x=0}^{L} |M - M_{ex}| dx}{\int_{x=0}^{L} |M_{ex}| dx} \tag{46}$$

where M_{ex} is the exact solution.

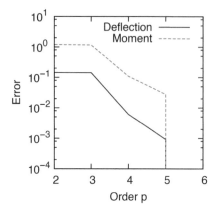

Figure 5 Plot of Error norms versus basis function order.

2.4 Approximation of unknown

We start the simulation with a basis function with the lowest order that would give non-zero results ($p = 2$), considering that the second derivative is involved in the solution. The errors in the solution for the approximation of w by Lagrange polynomials and B-splines are exactly the same:

$$\|\epsilon_w\| = 0.143; \quad \|\epsilon_M\| = 1.139 \tag{47}$$

As expected, since the basis functions of order 2 can only approximate a constant variation of w'' and therefore of the moment M we get quite a large error. One way of improving the solution is to increase the order of the basis functions.

2.5 P-refinement or order elevation

If we use Lagrange/Serendipity functions, the term used in the literature for increasing the function order is p-refinement and if we use B-splines it is order elevation. In this case both approximations yield the same result.

For the assumed quadratic variation of the loading q, the exact solution (w_{ex}) can be obtained with basis functions of order 6. We start with order 2 and elevate until the exact solution is obtained. Figure 4 shows the deflected shape and the bending moment diagrams for the different refinement stages.

The error norms as a function of order are plotted in Figure 5. It can be seen that there is no change in error from order 2 to 3. The reason for this can be seen in the bending moment diagram, which is approximated by a constant and linear variation.

2.6 H-refinement, the Finite Element Method

Another way of refining the solution when using Lagrange/Serendipity basis functions is h-refinement. In this method the beam is subdivided into finite elements as shown in Figure 6.

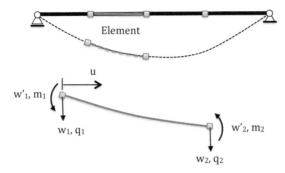

Figure 6 Discretization of beam into elements.

For each finite element the deflection is described with a local basis function. Compatibility and equilibrium conditions are enforced where elements join.

If we apply Lagrange polynomials then h-refinement will not work because the functions will only have a C^0 continuity where elements join but the solution requires at least a C^1 continuity along the whole length of the beam. The only way around this is the introduction of a rotational degree of freedom ($=$ the first derivative of w) and the introduction of basis functions that include this as parameter. The Hermite polynomials, introduced earlier, fulfill this requirement.

The approximation of the deflection w inside an element is now expressed in terms of local coordinate u as

$$w(u) = H_1(u) \cdot w_1 + H_2(u) \cdot w_1' + H_3(u) \cdot w_2 + H_4(u) \cdot w_2' \tag{48}$$

We denote the deflection and the rotation at node i as w_i and w_i'.
Rewriting Equation (48) in matrix notation we have

$$w(u) = \mathbf{H} \cdot \mathbf{w}_e \tag{49}$$

where

$$\mathbf{H} = \big(H_1(u),\, H_2(u),\, H_3(u),\, H_4(u) \big) \tag{50}$$

and

$$\mathbf{w}_e = \begin{pmatrix} w_1 \\ w_1' \\ w_2 \\ w_2' \end{pmatrix} \tag{51}$$

The second derivative of w is given by

$$w''(u) = \frac{d^2 H_1(u)}{dx^2} \cdot w_1 + \frac{d^2 H_2(u)}{dx^2} \cdot w_1' + \frac{d^2 H_3(u)}{dx^2} \cdot w_2 + \frac{d^2 H_4(u)}{dx^2} \cdot w_2' \tag{52}$$

and considering that $\frac{d^2u}{dx^2} = 0$:

$$\frac{d^2 H_i(u)}{dx^2} = \frac{d^2 H_i(u)}{du^2} \cdot \left(\frac{du}{dx}\right)^2 \tag{53}$$

where

$$\frac{du}{dx} = \frac{1}{L_e} \tag{54}$$

L_e is the length of the element.

In matrix notation we can write:

$$w''(u) = \mathbf{H}'' \cdot \mathbf{w}_e \tag{55}$$

where

$$\mathbf{H}'' = \left(\frac{d^2 H_1(u)}{dx^2}, \frac{d^2 H_2(u)}{dx^2}, \frac{d^2 H_3(u)}{dx^2}, \frac{d^2 H_4(u)}{dx^2}\right) \tag{56}$$

Computation of stiffness matrix and force vector Using the principle of virtual work we apply virtual displacements δw as well as virtual rotations δw'.

Applying a virtual displacement $δw_1$ for example, the virtual work done by the distributed force q is:

$$\delta W_e^q = \int_{x=0}^{L} q(H_1(u) \cdot \delta w_1) \cdot dx = \left(\int_{x=0}^{L} q H_1(u) \cdot dx\right) \cdot \delta w_1 \tag{57}$$

For the strain energy we get

$$\delta W_i = -\int_{x=0}^{L} E \cdot I \cdot (\mathbf{H}'' \cdot \mathbf{w}_e) \cdot \frac{\partial^2 H_1(u)}{\partial x^2} \cdot \delta w_1 \cdot dx \tag{58}$$

After substitutions and integrations (which we can perform analytically) we obtain the stiffness matrix of the beam element as

$$k_e = EI \cdot \begin{pmatrix} \dfrac{12}{L_e^3} & \dfrac{6}{L_e^2} & -\dfrac{12}{L_e^3} & \dfrac{6}{L_e^2} \\[3mm] \dfrac{6}{L_e^2} & \dfrac{4}{L_e} & -\dfrac{6}{L_e^2} & \dfrac{2}{L_e} \\[3mm] -\dfrac{12}{L_e^3} & -\dfrac{6}{L_e^2} & \dfrac{12}{L_e^3} & -\dfrac{6}{L_e^2} \\[3mm] \dfrac{6}{L_e^2} & \dfrac{2}{L_e} & -\dfrac{6}{L_e^2} & \dfrac{4}{L_e} \end{pmatrix} \tag{59}$$

an the force vector as

$$\mathbf{f}_e = \begin{pmatrix} q_1 \\ m_1 \\ q_2 \\ m_2 \end{pmatrix} = \begin{pmatrix} \int_{x=0}^{L_e} q \cdot H_1(\mathbf{u})dx \\ \int_{x=0}^{L_e} q \cdot H_2(\mathbf{u})dx \\ \int_{x=0}^{L_e} q \cdot H_3(\mathbf{u})dx \\ \int_{x=0}^{L_e} q \cdot H_4(\mathbf{u})dx \end{pmatrix} \tag{60}$$

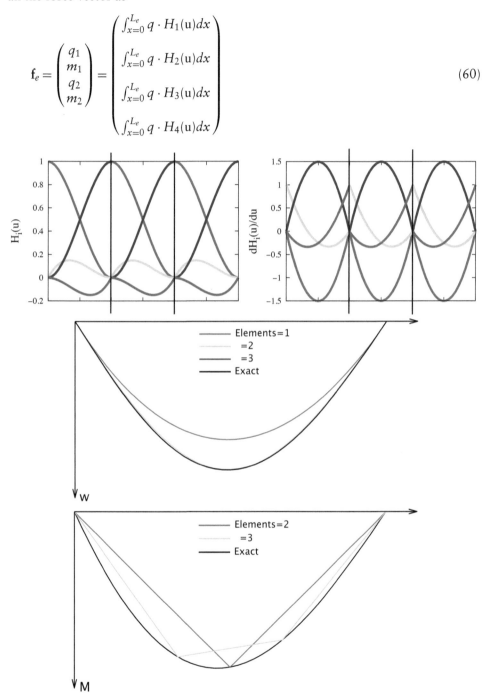

Figure 7 Plot of basis functions for the approximation of the unknown with Hermite polynomials and their first derivatives for subdivision into 3 Elements. Deflected shape and bending moment diagram for 3 refinement stages compared with the exact solution.

Assembly   After obtaining the stiffness matrices of all beam elements we assemble them. For this we use the equations of compatibility and equilibrium. Connecting end 2 of element 1 to end 1 of element 2 we have:

$$w_{2,1} = w_{1,2}; \quad w'_{2,1} = w'_{1,2}$$
$$q_{2,1} + q_{1,2} = Q; \quad m_{2,1} + m_{1,2} = M \tag{61}$$

where the first subscript denotes the node number and the second one the element number. The Equations (61) mean that the relevant stiffness coefficients and load of connected elements are added. We now investigate the convergence as a function of the number of elements.

In Figure 7 we show the unrefined results (with one element) and the results of the first two stages of refinement. Also shown are the basis functions for the discretization into 3 elements. We can see that the unrefined results are very inaccurate but that after the second stage of refinement (3 Elements) the deflected shapes are indistinguishable from the exact solution. It is quite a different picture for the bending moments. Without refinement the bending moment is actually zero. Although the moment is predicted exactly at the element nodes for the refined solutions, the variation within the element is in considerable error since the element is only able to represent a linear variation of moment.

2.7 Knot insertion, isogeometric method

Knot insertion is similar to the h-refinement just discussed. However, there is a subtle difference: No subdivision into elements is required. Since the basis functions retain a C^1 continuity (see Figure 8) there is no need to introduce rotational degrees of freedom. Unlike the previous case, we can even start with a function of order 2. The refinement is performed by repeatedly inserting equally spaced knots, i.e. the knot vector for the first refinement is $\Xi = (0\ 0\ 0\ 0.5\ 1\ 1\ 1)$ and the second refinement $\Xi = (0\ 0\ 0\ 0.33\ 0.66\ 1\ 1\ 1)$ and so on. As can be seen in Figure 8 the beam deflection is not as well approximated as with the cubic Hermite functions. This is not surprising as the second derivative of the functions is constant and therefore only a constant value of M can be achieved between knots. Better results can be achieved by increasing the order before inserting knots. This will be discussed next.

2.8 K-refinement

This type of refinement is unique to B-splines. It involves the elevation of the order before inserting the knots. This means that the knot vector for the first refinement is $\Xi = (0\ 0\ 0\ 0\ 0.5\ 1\ 1\ 1\ 1)$ and for the second refinement $\Xi = (0\ 0\ 0\ 0\ 0.33\ 0.66\ 1\ 1\ 1\ 1)$ and so on. Figure 9 shows the deflected shapes and the bending moment diagrams for k-refinement (order elevated to $p = 3$ and then knots inserted).

The convergence of the displacements is also shown in Figure 10. It can be seen that k-refinement produces the best results.

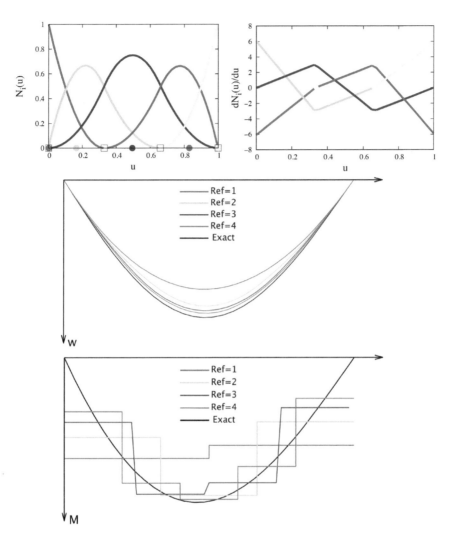

Figure 8 Basis functions of order p = 2 and their first derivatives after insertion of 2 knots, showing knots as squares and anchors as colored filled circles. Deflection and bending moment
diagrams for 4 refinement stages (Knot insertions).

2.9 Summary and conclusions

The main aim of this section was to introduce readers to the process of simulation but in
particular the objective was to show the difference between simulation with Lagrange
functions and B-splines. This was made easier by using a simple one dimensional
example. The superior properties of B-splines with respect to control of continuity
and refinement strategies should have become clear. The use of B-splines allowed a

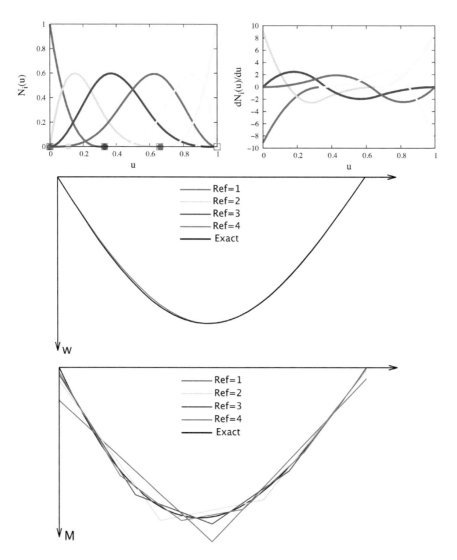

Figure 9 Plot of basis functions for refinement stage 2 (order elevation to p = 3 and 2 knot insertions). Beam deflection and bending moment diagrams for 4 k-refinement stages.

simulation of the beam without discretization into elements and especially without the introduction of rotational degrees of freedom.

Notation Before proceeding to more realistic problems a remark on the notation used is appropriate: To comply with published literature we use u, v for the local coordinates for B-splines/NURBS. This should not be confused with **u** (bold) for the vector of displacements with the components u_x, u_y, u_z.

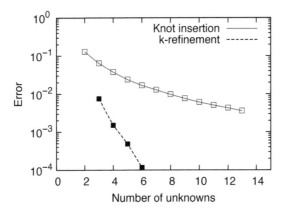

Figure 10 Convergence of displacements for the case of knot insertion and k-refinement.

3 2-D SIMULATION, PLANE STRESS AND PLANE STRAIN

We now extend the simulation to two dimensions. Examples are the analysis of thin plates with in plane loading assuming *plane stress* conditions or underground excavations under *plane strain* conditions. In the first case it is assumed that the stress in the direction perpendicular to the plane is zero in the second that the strain perpendicular to the plane is zero. The geometry of the problem can be described by NURBS:

$$\mathbf{x} = \sum_{i=1}^{I_g} N_i^g(\mathrm{u}, \mathrm{v}) \cdot \mathbf{x}_i \tag{62}$$

where $\mathbf{x}_i$ are the coordinates of control points and the superscript g denotes that the basis functions may be different from those describing the displacements.

The strain tensor is of dimension 2×2. For convenience we put the strain components into a pseudo-vector using the Voight notation[5].

$$\boldsymbol{\epsilon} = \begin{pmatrix} \epsilon_x \\ \epsilon_y \\ \tau_{xy} \end{pmatrix} = \begin{pmatrix} \dfrac{\partial u_x}{\partial x} \\[2mm] \dfrac{\partial u_y}{\partial y} \\[2mm] \dfrac{\partial u_x}{\partial y} + \dfrac{\partial u_y}{\partial x} \end{pmatrix} \tag{63}$$

The displacement vector $\mathbf{u}$ has 2 components and is approximated by B-splines $N_i(\mathrm{u}, \mathrm{v})$ as:

$$\mathbf{u} = \sum_{i=1}^{I} N_i(\mathrm{u}, \mathrm{v}) \cdot \mathbf{u}_i \tag{64}$$

---

[5]Voigt notation is a way to represent a symmetric tensor by a one column matrix (pseudo vector).

The derivatives of the displacement vector are given by

$$\frac{\partial u}{\partial x} = \sum_{i=1}^{I} \frac{\partial N_i(u, v)}{\partial x} \cdot u_i; \quad \frac{\partial u}{\partial y} = \sum_{i=1}^{I} \frac{\partial N_i(u, v)}{\partial y} \cdot u_i \tag{65}$$

The global derivatives of the shape functions can be obtained via the chain rule of differentiation:

$$\frac{\partial N_i(u, v)}{\partial u} = \frac{\partial N_i}{\partial x} \cdot \frac{\partial x}{\partial u} + \frac{\partial N_i}{\partial y} \cdot \frac{\partial y}{\partial u} \tag{66}$$

$$\frac{\partial N_i(u, v)}{\partial v} = \frac{\partial N_i}{\partial x} \cdot \frac{\partial x}{\partial v} + \frac{\partial N_i}{\partial y} \cdot \frac{\partial y}{\partial v} \tag{67}$$

or in matrix form

$$\mathbf{N}_{,u} = \mathbf{J} \cdot \mathbf{N}_{,x} \tag{68}$$

where

$$\mathbf{N}_{,u} = \begin{pmatrix} \dfrac{\partial N_i}{\partial u} \\[2mm] \dfrac{\partial N_i}{\partial v} \end{pmatrix}; \quad \mathbf{N}_{,x} = \begin{pmatrix} \dfrac{\partial N_i}{\partial x} \\[2mm] \dfrac{\partial N_i}{\partial y} \end{pmatrix} \tag{69}$$

and the Jacobi matrix is given by

$$\mathbf{J} = \begin{pmatrix} \dfrac{\partial x}{\partial u} & \dfrac{\partial y}{\partial u} \\[2mm] \dfrac{\partial x}{\partial v} & \dfrac{\partial y}{\partial v} \end{pmatrix} \tag{70}$$

with

$$\frac{\partial \mathbf{x}}{\partial u} = \sum_{i=1}^{I_g} \frac{\partial N_i^g(u, v)}{\partial u} \cdot \mathbf{x}_i; \quad \frac{\partial \mathbf{x}}{\partial v} = \sum_{i=1}^{I_g} \frac{\partial N_i^g(u, v)}{\partial v} \cdot \mathbf{x}_i \tag{71}$$

The determinate of the Jacobi matrix ($J = |\mathbf{J}|$) is the Jacobian introduced earlier.
The global derivatives of the basis functions are obtained by using the inverse of $\mathbf{J}$:

$$\mathbf{N}_{,x} = \mathbf{J}^{-1} \cdot \mathbf{N}_{,u} \tag{72}$$

The stresses are related to strains via the elasticity matrix

$$\boldsymbol{\sigma} = \mathbf{C} \cdot \boldsymbol{\epsilon} \tag{73}$$

where for plane stress:

$$C = \frac{E}{1 - v^2} \begin{pmatrix} 1 & v & 0 \\ v & 1 & 0 \\ 0 & 0 & 1 - v \end{pmatrix} \tag{74}$$

and for plane strain:

$$C = \frac{E}{(1 + v)(1 - v)} \begin{pmatrix} 1 - v & v & 0 \\ v & 1 - v & 0 \\ 0 & 0 & 1 - 2v \end{pmatrix} \tag{75}$$

E is the modulus of elasticity and v is the Poisson's ratio.

After substitution of (65) into (63) we obtain a relationship between strains and displacements:

$$\epsilon = \sum_{i=1}^{I} \mathbf{B}_i \cdot \mathbf{u}_i \tag{76}$$

where

$$\mathbf{B}_i = \begin{pmatrix} \dfrac{\partial N_i}{\partial x} & 0 \\ 0 & \dfrac{\partial N_i}{\partial y} \\ \dfrac{\partial N_i}{\partial y} & \dfrac{\partial N_i}{\partial x} \end{pmatrix} \tag{77}$$

Using the principle of virtual work, as explained previously, we apply virtual displacements $\delta\mathbf{u}_j$ and compute the internal virtual work

$$\delta W_{\text{int}} = -\int_V \sigma^T \delta\epsilon \cdot dV \tag{78}$$

After substitution of (76) and (73) we obtain

$$\delta W_{\text{int}} = -\int_V \left[C \sum_{i=1}^{I} \mathbf{B}_i \cdot \mathbf{u}_i \right]^T \mathbf{B}_j \cdot \delta\mathbf{u}_j \cdot dV \tag{79}$$

For a distributed body force $\mathbf{b}$ (force per unit volume) the external virtual work is

$$\delta W_{\text{ext}} = \int_V \mathbf{b} \cdot \delta\mathbf{u}_j \cdot dV \tag{80}$$

For the numerical integration we need to map the u, v coordinates to ξ, η coordinates that range for -1 to 1:

$$u = 0.5(\xi + 1); \quad v = 0.5(\eta + 1) \tag{81}$$

The Jacobian of this transformation is $J_u = 0.5 \times 0.5 = 0.25$. The submatrices of the element stiffness matrix are given by:

$$\mathbf{k}_{ij}^e = d \cdot \int_{-1}^{1} \int_{-1}^{1} \mathbf{B}_i^T \cdot \mathbf{C} \cdot \mathbf{B}_j \cdot J \cdot J_u \cdot d\xi \cdot d\eta \tag{82}$$

and the sub-vectors of the force vector are:

$$\mathbf{f}_i^e = d \cdot \int_{-1}^{1} \int_{-1}^{1} \mathbf{b} \cdot N_i(\mathrm{u}, \mathrm{v}) \cdot J \cdot J_u \cdot d\xi \cdot d\eta \tag{83}$$

where d is the thickness of the plate.

For the evaluation of the integrals we use an $M \times N$ Gauss Quadrature rule for the computation of the stiffness:

$$\mathbf{k}_{ij}^e = d \cdot \sum_{m=1}^{M} \sum_{n=1}^{N} \mathbf{B_i}^T(\xi_n, \eta_m) \cdot \mathbf{C} \cdot \mathbf{B}_j(\xi_n, \eta_m) \cdot J \cdot J_u \cdot W_n \cdot W_m \tag{84}$$

and for the forces:

$$\mathbf{f}_i^e = d \cdot \sum_{m=1}^{M} \sum_{n=1}^{N} \mathbf{b} \cdot N_i(\xi_n, \eta_m) \cdot J \cdot J_u \cdot W_n \cdot W_m \tag{85}$$

The number of Gauss points required depends on the order of the functions used. Traditionally, implementations of the FEM have used linear or quadratic basis functions, so the number of Gauss points can be limited to 3. Indeed, it has been found that an under-integration may be beneficial as meshes with a coarse discretization always tend to be too stiff. For the implementation with NURBS the order of the basis functions can be quite high and therefore care has to be taken that the number of integration points is sufficient for an accurate integration. One indication that the number is insufficient, is that the stiffness matrix becomes singular or nearly singular.

3.1 Boundary Conditions (BC)

There are three types of boundary conditions:

- Displacements are known
- Forces or tractions are known
- Forces or tractions can be computed using geometrical information

Dirichlet BC   The first type is also known as Dirichlet boundary condition[6].
There are two ways in which this BC can be implemented:

- Strong form: If the displacement component is specified it means that its value is known and need not be computed. If the component is zero, we simply delete the row and column associated with it in the stiffness matrix. If the specified displacement is not zero then we move the associated stiffness coefficients to the right hand side and multiply them with the known displacement (see [1] for details of implementation).
- Weak form: A zero value of the boundary condition can be implemented by adding a large value to the stiffness coefficient associated with the displacement. This is the same as connecting a spring with a large stiffness and will result in the value of the displacement to be nearly zero (depending on the value of the stiffness added).

Neuman BC   The second type is also known as Neuman boundary condition[7].
They are implemented in the same way as shown for the body force except that the integral is taken along the boundary and not the volume. For example for a distributed loading q along the boundary $v = v_q$ we have:

$$f_i^q = d \cdot \int_{-1}^{1} \mathbf{q} \cdot N_i(\mathbf{u}, v_q) \cdot \bar{J} \cdot 0.5 \cdot d\xi \tag{86}$$

where

$$\bar{J} = \sqrt{\frac{\partial x}{\partial \mathbf{u}}^2 + \frac{\partial y}{\partial \mathbf{u}}^2} \tag{87}$$

Robin BC   The third type is known as Robin boundary condition[8].
Applied to heat flow, the component of the flow q normal to the boundary (*t*) is computed as:

$$t = \mathbf{n} \cdot \mathbf{q} \tag{88}$$

where **n** is a unit vector normal to the boundary.

---

[6]The Dirichlet (or first-type) boundary condition is a type of boundary condition, named after Peter Gustav Lejeune Dirichlet (1805 to 1859). When imposed on an ordinary or a partial differential equation, it specifies the values that a solution needs to take on along the boundary of the domain.

[7]The Neumann (or second-type) boundary condition is a type of boundary condition, named after Carl Neumann. When imposed on an ordinary or a partial differential equation, it specifies the values that the derivative of a solution is to take on the boundary of the domain.

[8]This is a type of boundary condition, named after Victor Gustave Robin (1855 to 1897) even though his original contribution is disputed by some. When imposed on an ordinary or a partial differential equation, it is a specification of a linear combination of the values of a function and the values of its derivative on the boundary of the domain.

Applied to an excavation problem where the excavation traction vector (t) is computed by

$$t = n \cdot \sigma_0 \qquad (89)$$

where σ_0 is a pseudo vector of virgin stress.

3.2 Using one NURBS patch

In contrast to the classical FEM our aim is to get away from the need to generate a mesh. This is possible because, as we have seen in the previous stages, NURBS can describe complex geometries with one patch and few parameters.

If the geometry can be described by one NURBS patch then the implementation is straight forward. The only consideration when evaluating the integrals (84) and (85) is with respect to the number of Gauss points required depending on the order of the B-splines. If the basis functions span only part of the parameter space, then the integration region has to be subdivided into subregions. The subregions are defined by the knot spans.

For example for the Knot vector $\Xi = (0, 0, 0, 0.3, 0.6, 1, 1, 1)$ the subdivision lines are placed at 0.3 and 0.6. For subregion k the relationship between u, v and ξ, η coordinates is given by:

$$u = \frac{s_u^k}{2} (\xi + 1) + u_s^k \qquad (90)$$

$$v = \frac{s_v^k}{2} (\eta + 1) + v_s^k$$

where s_u^k, s_v^k specify the extent of the subregion and u_s^k, v_s^k are its starting coordinates.

The Jacobian of this transformation is $J_u = 0.5 \cdot s_u^k \cdot 0.5 \cdot s_v^k$.

The numerical integration for the stiffness and force terms is given by:

$$k_{ij} = d \cdot \sum_{k=1}^{K} \sum_{m=1}^{M} \sum_{n=1}^{N} B_i^T(u_n, v_m) \cdot C \cdot B_j(u_n, v_m) \cdot J \cdot J_u \cdot W_n \cdot W_m \qquad (91)$$

and

$$f_i = d \cdot \sum_{k=1}^{K} \sum_{m=1}^{M} \sum_{n=1}^{N} b \cdot N_i(u_n, v_m) \cdot J \cdot J_u \cdot W_n \cdot W_m \qquad (92)$$

where M and N are the number of Gauss points and K is the number of subregions.

For the implementation of Dirichlet BC's we have to consider that nodal points do not exist and therefore a different approach than with the isoparametric FEM has to be taken. We recall that the displacements are defined by parameter values and this means that they are associated with anchors rather than nodal points. Therefore the entries in the stiffness matrix refer to values of the parameters at the anchors. To define a zero value of the displacement we adopt the weak form as explained above and add a large value to the diagonal coefficient of the stiffness matrix associated with the degree of freedom. Depending on the value inserted in the stiffness matrix, the resulting value of displacement will not be exactly zero but a very small number.

3.3 Comparison with classical FEM

In the classical FEM (for details see [4]) we subdivide the domain into finite elements and using either Lagrange polynomials or Serendipity functions. For each element we have

$$\mathbf{x}^e = \sum_{i=1}^{I} N_i(\xi, \eta) \cdot \mathbf{x}_i^e \tag{93}$$

where $\mathbf{x}_i^e$ are nodal point coordinates and

$$\mathbf{u}^e = \sum_{i=1}^{I} N_i(\xi, \eta) \cdot \mathbf{u}_i^e \tag{94}$$

where $\mathbf{u}_i^e$ are nodal point displacements.

The superscript e indicates that the approximation is only valid within element e. Usually the same functions are used for the description of the geometry of an element and the variation of the unknown, so we have dropped the superscript g. These elements are also known as *isoparametric elements*.

After computing the stiffness matrices and force vectors for all elements we assemble them into the global system using compatibility and equilibrium conditions between the elements. C^0 compatibility between element e_1 and e_2 is ensured if the displacements of the nodes where the elements connect and the basis functions used for their approximations are the same.

Compatibility between node i of element e_1 and node j of element e_2 is ensured by

$$\mathbf{u}_i^{e1} = \mathbf{u}_j^{e2} = \mathbf{u}_{inci(i)^{e1}} \tag{95}$$

Equilibrium is assured by the condition

$$\mathbf{f}_i^{e1} + \mathbf{f}_j^{e2} = \mathbf{F}_{inci(i)^{e1}} \tag{96}$$

where $\mathbf{F}_{inci(i)}$ is a nodal point force vector and $inci(i)^{e1}$ ($=inci(j)^{e2}$) is the global number of the local node i of element $e1$ also referred to as *incidence*. Equations (95) and (96) mean that corresponding stiffness coefficients are assembled as follows:

$$\mathbf{K}_{inic(i)^e, inci(j)^e} = \mathbf{K}_{inic(i)^e, inci(j)^e} + \mathbf{k}_{i,j}^e \tag{97}$$

For the computation of the stiffness matrix, equations (84) and (85) are used directly and no transformation from u, v coordinates is necessary and $J_u = 1$. For the imposition of Dirichlet boundary conditions we use the strong form, i.e. we specify displacement values at the nodal points.

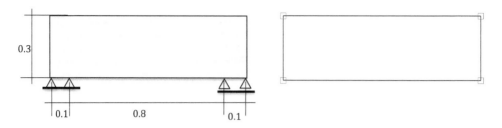

Figure 11 Geometry of the problem and description of the geometry with one NURBS patch.

3.4 Example

First we present a simple example in order to explain the differences between the classical FEM and NURBS based analysis. It is a plate supported at two edges and subjected to gravity loading.

The input data are:

- Length = 1 m, height = 0.3 m, thickness = 1 m
- E = 100, Poissons ratio = 0
- Specific weight = 1

For the analysis with NURBS we first define the geometry using linear basis functions and Knot vectors $\Xi_u^g = \Xi_v^g = (0, 0, 1, 1)$.

The coordinates and weights of the control points are given by:

```
0  0    0  1
1  0    0  1
0  0.3  0  1
1  0.3  0  1
```

Figure 11 shows the geometry of the problem and the discretization with one NURBS patch.

The first choice of basis functions for the approximation of the unknown must include anchors at the location where the Dirichlet BC have to be enforced. This means that we need to introduce basis functions that have anchors at the location of the supports before refinement. The functions can be defined by the following knot vectors: $\Xi_u = (0, 0, 0.1, 0.9, 1, 1)$; $\Xi_v = (0, 0, 1, 1)$. Figure 12 shows the basis functions and the associated anchors before refinement and after the first stage of order elevation from p = 1 to p = 2 prior to a k-refinement (i.e. prior to inserting knots).

The anchors where the restraint condition was enforced are depicted. Figure 13 shows the displaced shape and the anchors for the first stage of k-refinement (one knot inserted in each direction). Also shown is the subdivision of the integration region into subregions, to take into account the limited span of the basis functions after knot insertion. As a comparison we show the displaced finite element mesh in Figure 13.

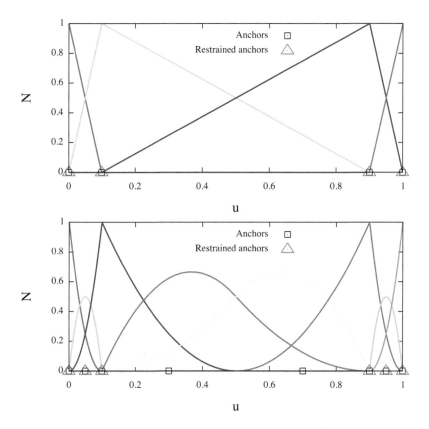

Figure 12   Plot showing the basis functions and the location of Anchors along the bottom of the plate
for the case with no refinement and for the first stage of k-refinement (order elevation to
$p = 2$ and one Knot insertion). Anchors, where boundary conditions have been enforced
are shown as triangles.

To test the convergence of the methods we study the change in displacement as
we implement the different refinement strategies. For the FEM we used h-refinement,
i.e. we have increased the number of elements and this is compared with k-refinement.
We plot the convergence of the maximum displacement in Figure 13. It can be seen
that convergence is much faster for the NURBS based simulation than for the Finite
Element Method.

3.5 Multiple NURBS patches

In some cases it is convenient to use more than one NURBS patch to describe the
geometry. Since there are no nodal points, compatibility conditions have to be defined
in a different way. C^0 compatibility between patches is ensured if the location of the
anchors match along the line where patches connect. These conditions can be enforced
in the same way as for the conventional FEM, i.e. by using incidences for the assembly
of coefficients, except that indices refer now to anchors instead of nodal points.

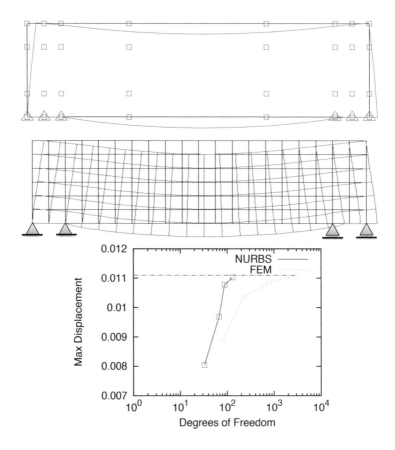

Figure 13   Plot showing deformed shape and location of Anchors for first stage of k-refinement (order elevation to p = 2 and one Knot insertion). Also shown is the subdivision into integration regions. Displaced FEM mesh and convergence of maximum displacement.

To explain this consider the plate with a hole in Figure 14 subjected to a tensile stress T, where we analyze one quarter with symmetry boundary conditions. The following input data was used:

- Length $= 5$, height $= 5$, thickness $= 1$
- $E = 10^5$, Poissons ratio $= 0.3$
- Tensile Stress $= 1$
- $R = 1$

This problem has been popular for convergence studies (see [2]) since there is an exact solution for a hole in an infinite domain. To be able to use the exact solution for the infinite space with a finite space problem, we have to ensure that the boundary conditions for both problems are the same. Therefore, the traction computed from the

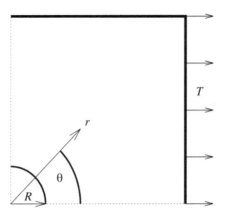

Figure 14 Plate with a hole: problem definition.

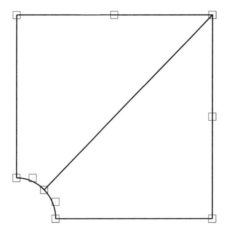

Figure 15 Geometry definition with two NURBS patches.

exact solution are applied at the boundaries. The exact solution for the radial stress (σ_{rr}), tangential stress ($\sigma_{\theta\theta}$) and the associated shear stress ($\sigma_{r\theta}$) is given by:

$$\sigma_{rr}(r,\theta) = \frac{T}{2} \cdot \left\{ 1 - \frac{R^2}{r^2} + \left(1 - 4\frac{R^2}{r^2} - 3\frac{R^4}{r^4} \right) \cos 2\theta \right\}$$

$$\sigma_{\theta\theta}(r,\theta) = \frac{T}{2} \cdot \left\{ 1 + \frac{R^2}{r^2} - \left(1 + 3\frac{R^4}{r^4} \right) \cos 2\theta \right\} \tag{98}$$

$$\sigma_{r\theta}(r,\theta) = -\frac{T}{2} \cdot \left(1 + 2\frac{R^2}{r^2} - 3\frac{R^4}{r^4} \right) \sin 2\theta$$

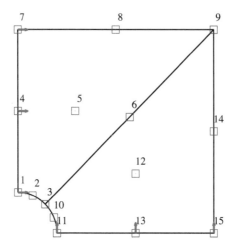

Figure 16 Approximation of displacements, showing location of anchors. Red arrows show restrained anchors.

We start with the description of the geometry with 2 NURBS patches as shown in Figure 15. The knot vectors are given by:

```
0 0 0 1 1 1
0 0 1 1

0 0 0 1 1 1
0 0 1 1
```

and the coordinates/weights of control points are:

```
0 1 0 1
0.4142 1 0 0.9238
0.707 0.707 0 1
0 5 0 1
2.5 5 0 1
5 5 0 1

0.707 0.707 0 1
1 0.4142 0 0.9238
1 0 0 1
5 5 0 1
5 2.5 0 1
5 0 0 1
```

We refine the approximation of the unknown by order elevation and knot insertion. Figure 16 shows the anchors of the basis functions for order elevating in the v direction from $p = 1$ to $p = 2$. We number the anchors sequentially making sure to assign a unique number to the anchors on the interface between the patches. Anchors that are restrained to simulate symmetry conditions are specified. For the assembly process we need to define the connectivity between the patches via an incidence vector.

The incidence vectors for patches 1 and 2 are:

$$inci(j)^1 = \begin{pmatrix} 1 & 2 & 3 & 4 & 5 & 6 & 7 & 8 & 9 \end{pmatrix}$$

$$inci(j)^2 = \begin{pmatrix} 3 & 10 & 11 & 6 & 12 & 13 & 9 & 14 & 15 \end{pmatrix}$$

The analysis proceeds as before with the computation of the stiffness matrix and the load vector of the patch using Equations (91) and (92) and the assembly of the stiffness matrices using (97). To refine the solution we can use the three refinement strategies available.

3.6 Bezièr elements

Although the aim throughout this book is to avoid mesh generation we present here for completeness a method known as *Bezièr elements*. In this method we mimic the classical Finite Element method. First we insert as many knots as necessary to make the basis functions only C^0 continuous between knot spans.

For example if we insert two knots in u and v directions we change the knot vectors to

```
0 0 0 0.5 0.5 1 1 1
0 0 0 0.5 0.5 1 1 1

0 0 0 0.5 0.5 1 1 1
0 0 0 0.5 0.5 1 1 1
```

In Figure 17 we show the traces of the basis functions along $u = 0$ and $v = 1$. It can be seen that since the basis functions have a C^0 continuity at the borders of the subregions of integration, these can be considered separately just like finite elements.

Instead of a global knot vector we use a local knot vector for each element, i.e.

```
0 0 0 1 1 1
0 0 0 1 1 1
```

Figure 17 bottom shows meshes generated. All edge nodes are interpolatory and therefore the assembly process follows the standard FEM procedures. The main difference to classical FEM is that the basis functions differ form the Serendipity functions and that the geometrical values such as location of Gauss points and Jacobi matrix are computed using NURBS technology. Therefore the main advantages of the isogeometric method can be retained.

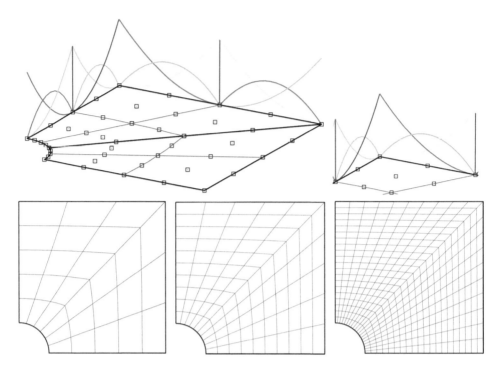

Figure 17 Top: Traces of basis functions prior to Bezièr extraction and extracted Bezièr element. Below: Meshes of Bezièr elements.

3.7 Trimmed **NURBS** patches

Another possibility to analyze this problem is to use trimmed patches. To describe the problem geometry with a trimmed NURBS patch we start with a simple square patch and two trimming curves as shown in Figure 18 top. Because one of the trimming curves has only a C^0 continuity, care has to be taken when choosing the basis functions for the description of the unknown before refinement and the regions used for the integration. In Figure 18 bottom we show that in this case the integration region is split into two and the continuity of the basis functions is changed to match the one of the mapping function.

In this Figure we also show the trace of the basis functions for the approximation of the unknown along $t = 1$ in the local s, t coordinate system and in the global x, y coordinate system. The knot vectors of the trimming curves are given by:

0 0 0 1 1 1
0 0 0.5 1 1

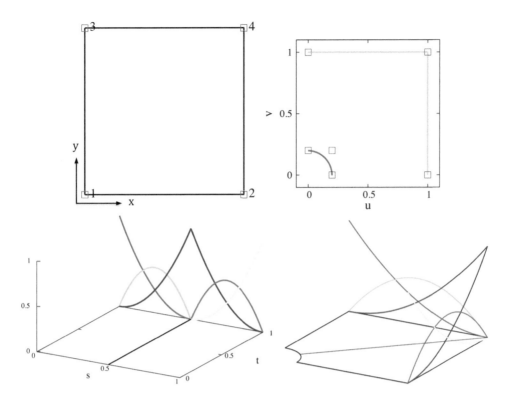

Figure 18 Geometry definition of plate with a hole: Top left: untrimmed NURBS surface in global *x, y* system and Top right: trimming curves in local u, v system. Bottom: trace of basis functions along t = 1 in the local s, t and in the global *x, y* coordinate system.

The coordinates of the control points and weights of the trimming curves are:

```
0 0.2  0 1
0.2 0.2  0 0.707
0.2 0  0 1

0 1 0 1
1 1 0 1
1 0 0 1
```

The implementation is the same as for the single patch, but as explained in Stage 3, we now have to include an additional mapping step from s, t to the u, v map before mapping to the Cartesian x, y, z system.

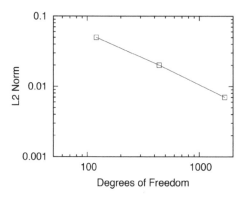

Figure 19 Convergence for plate with hole.

3.8 Convergence test

For convergence tests it is common to use the L2 error in the strain energy norm, defined as:

$$L2 = \sqrt{\frac{\int_\Omega (\boldsymbol{\epsilon}_e - \boldsymbol{\epsilon})^T \cdot (\boldsymbol{\sigma}_e - \boldsymbol{\sigma})^T d\Omega}{\int_\Omega \boldsymbol{\epsilon}_e^T \boldsymbol{\sigma}_e \, d\Omega}} \tag{99}$$

where $\boldsymbol{\epsilon}_e$ and $\boldsymbol{\sigma}_e$ are the exact values of strain and stress.

We show the convergence of the L2 Norm for a k-refinement with order 2 in Figure 19. All approaches discussed here give the same result.

4 SUMMARY AND CONCLUSIONS

In this stage we have introduced the steps in numerical simulation. The main aim was to show the subtle differences between the conventional FEM approach and the one using NURBS. We started with a simple one-dimensional example, which served well to demonstrate the additional flexibility gained by using B-splines for the approximation of the unknown. Proceeding to two dimensions, it was shown that mesh generation can be avoided if the geometry of the problem is described by NURBS patches. On practical examples it was shown that using B-splines for the approximation of the unknown results in better convergence rates, i.e. results can be obtained with fewer degrees of freedom.

In the next stage we venture into three dimensional space but restrict ourselves to surfaces, thus still taking advantage of the CAD philosophy.

BIBLIOGRAPHY

[1] B.M. Irons and S. Ahmad. *Techniques of Finite Elements.* Halstead Press, 1980.
[2] J. Austin Cottrell, Thomas J.R. Hughes and Yuri Bazilevs. *Isogeometric Analysis: Toward Integration of CAD and FEA.* John Wiley & Sons, Chichester, England, 2009.

[3]   W. Ritz. Über eine neue Methode zur Lösung gewisser Variations probleme der mathematischen Physik. *Journal für die Reine und Angewandte Mathematik*, 135:1 to 61, 1909.

[4]   I.M. Smith, D.V. Griffiths and L. Margetts. *Programming the Finite Element Method.* Wiley, 2013.

[5]   S.P. Timoshenko and J.N. Goodier. *Theory of Elasticity.* McGraw-Hill, New York, NY, USA, 1970.

Chapter 6

Stage 5: Plates and shells

There is nothing more practical than a good theory

I. Kant

where we venture into three-dimensional space but stay in two local dimensions.

I KIRCHHOFF PLATE

The Kirchhoff[1]–Love[2] theory is an extension of the Bernoulli beam theory to thin plates and shells. The theory was developed in 1888 by Love using assumptions proposed by Kirchhoff. It is assumed that the mid-surface can be used to represent the three-

Figure I Gustav Robert Kirchhoff.

---

[1]Gustav Robert Kirchhoff (1824 to 1887) was a German physicist who contributed to the fundamental understanding of electrical circuits, spectroscopy, and the emission of black-body radiation by heated objects.
[2]Augustus Edward Hough Love (1863 to 1940), often known as A. E. H. Love, was a mathematician famous for his work on the mathematical theory of elasticity.

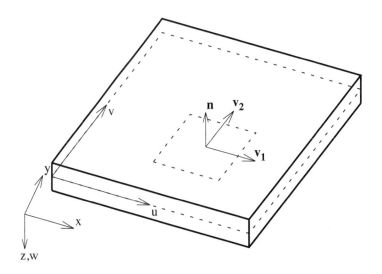

Figure 2 Geometry definition of Kirchhoff plate.

dimensional plate or shell. The following kinematic assumptions are made in this theory [5]:

- straight lines normal to the mid-surface remain straight and normal to the mid-surface after deformation
- the thickness of the plate does not change during a deformation.

This means that thin plates and shells can be represented by surfaces with two local coordinates.

1.1   Plates

We explain this theory first on a planar NURBS surface with loading perpendicular to the plane (plate) and extend it later to a curved NURBS surface (shell). As shown in Figure 2, the plate of thickness d is represented by a planar NURBS patch with local coordinates u and v at mid thickness. The geometry is defined by:

$$\mathbf{x} = \sum_{n=1}^{N} R_n(\mathrm{u}, \mathrm{v}) \mathbf{x}_n \qquad (1)$$

First we establish a local coordinate system u, v in the directions of vectors $\mathbf{v}_1$, $\mathbf{v}_2$. The vectors are given by

$$\mathbf{v}_1 = \begin{pmatrix} v_{1x} \\ v_{1y} \\ v_{1z} \end{pmatrix} = \frac{\partial \mathbf{x}}{\partial \mathrm{u}} ; \quad \mathbf{v}_2 = \begin{pmatrix} v_{2x} \\ v_{2y} \\ v_{2z} \end{pmatrix} = \frac{\partial \mathbf{x}}{\partial \mathrm{v}} \qquad (2)$$

We construct a vector normal to these two by taking the x-product

$$\mathbf{N} = \mathbf{v}_1 \times \mathbf{v}_2 \tag{3}$$

and normalize it

$$\mathbf{n} = \begin{pmatrix} n_x \\ n_y \\ n_z \end{pmatrix} = \frac{\mathbf{N}}{J} \tag{4}$$

where $J = \sqrt{n_x^2 + n_y^2 + n_z^2}$ is the Jacobian.

Assuming the plate is defined in the xy plane, the normal vector is $\mathbf{n} = (0\ 0\ 1)$ but we will generalize this for shells later. There are now 3 strain components (normal strains in u and v directions and a shear strain).

Following the derivation of the Bernoulli beam the strains are in Voigt notation:

$$\begin{pmatrix} \epsilon_u \\ \epsilon_v \\ \gamma_{uv} \end{pmatrix} = \begin{pmatrix} \kappa_u \cdot \zeta \cdot \dfrac{d}{2} \\[2ex] \kappa_v \cdot \zeta \cdot \dfrac{d}{2} \\[2ex] \kappa_{uv} \cdot \zeta \cdot \dfrac{d}{2} \end{pmatrix} \tag{5}$$

where ζ is a local coordinate perpendicular to the plate that ranges from -1 to 1.

The local curvatures are given by:

$$\kappa_u = \frac{\partial^2 w}{\partial u^2}; \quad \kappa_v = \frac{\partial^2 w}{\partial v^2}; \quad \kappa_{uv} = \frac{\partial^2 w}{\partial u \partial v} + \frac{\partial^2 w}{\partial v \partial u} = 2 \cdot \frac{\partial^2 w}{\partial v \partial u} \tag{6}$$

We approximate the deflection w with NURBS:

$$w = \sum_{n=1}^{N_d} R_n^d(u, v) w_n \tag{7}$$

where w_n are parameter values and the superscript d indicates that different functions to the ones for the description of the geometry will be used. The local curvatures are then computed by

$$\kappa_u = \sum_{n=1}^{N_d} \frac{\partial^2 R_n^d(u, v)}{\partial u^2} w_n$$

$$\kappa_v = \sum_{n=1}^{N_d} \frac{\partial^2 R_n^d(u, v)}{\partial v^2} w_n \tag{8}$$

$$\kappa_{uv} = 2 \cdot \sum_{n=1}^{N_d} \frac{\partial^2 R_n^d(u, v)}{\partial u \partial v} w_n$$

Care has to be taken to account for the fact that the coordinate system u and v is not a Cartesian system and in general will not be orthogonal (i.e. the plate may be skew). Therefore the local derivatives can not be directly used for computing the internal work. The conversion of local curvatures to global ones for the general case where the local axes are not orthogonal and aligned to the global axes can be performed using tensor analysis. The derivation presented here follows closely the one presented in [3], [2] and [1].

We introduce a metric tensor[3]:

$$A = \begin{pmatrix} A_{11} & A_{12} \\ A_{21} & A_{22} \end{pmatrix} \tag{9}$$

with $A_{ij} = \mathbf{v}_i \cdot \mathbf{v}_j$.

The inverse metric tensor is:

$$a = A^{-1} \tag{10}$$

For example, the metric tensor for the Bernoulli beam, discussed earlier, can be obtained by setting $\mathbf{v}_1 = (\frac{\partial x}{\partial u}\ 0\ 0)$ and $\mathbf{v}_2 = (0\ 0\ 0)$. In this case the tensor reverts to a scalar and we have:

$$A = \left(\frac{dx}{du}\right)^2; \quad a = \left(\frac{du}{dx}\right)^2 \tag{11}$$

which results in the multiplier for the local curvatures, that we obtained from applying the chain rule, considering that the curvature of the geometry was zero.

The stiffness matrix of the plate can be obtained using the principle of virtual work as explained for the beam. The coefficients of the stiffness matrix are given by:

$$k_{ij} = \frac{d^3}{12} \int_{\xi=-1}^{1} \int_{\eta=-1}^{1} \mathbf{B}_i^T \cdot \mathbf{C} \cdot \mathbf{B}_j \cdot J \cdot 0.25 \cdot d\xi\, d\eta \tag{12}$$

where

$$\mathbf{B}_i = \begin{pmatrix} \dfrac{\partial^2 R_i^d}{\partial u^2} \\[2ex] \dfrac{\partial^2 R_i^d}{\partial v^2} \\[2ex] 2 \cdot \dfrac{\partial^2 R_i^d}{\partial u \partial v} \end{pmatrix} \tag{13}$$

---

[3]In the mathematical field of differential geometry, a metric tensor is a function defined on a surface which takes as input a pair of tangent vectors $\mathbf{v}_1$ and $\mathbf{v}_2$ and produces $g(\mathbf{v}_1, \mathbf{v}_2)$ in a way that generalizes many of the familiar properties of the dot product of vectors in Euclidean space.

All transformations between the local and global coordinate systems are incorporated in the material matrix, which is given by:

$$C = \frac{E}{1-v^2} \begin{pmatrix} a_{11}^2 & va_{11}a_{22} + (1-v)a_{12}^2 & a_{11}a_{12} \\ va_{11}a_{12} + (1-v)a_{12}^2 & a_{22}^2 & a_{12}a_{22} \\ a_{11}a_{12} & a_{12}a_{22} & \frac{1}{2}[(1-v)a_{11}a_{22} + (1+v)a_{12}^2] \end{pmatrix} \quad (14)$$

For the special case of a plate, where the local axes u and v are parallel to the global axes x, y we have:

$$A = \begin{pmatrix} \left(\frac{\partial x}{\partial u}\right)^2 & 0 \\ 0 & \left(\frac{\partial y}{\partial v}\right)^2 \end{pmatrix} \quad (15)$$

and

$$a = \begin{pmatrix} \left(\frac{\partial x}{\partial u}\right)^{-2} & 0 \\ 0 & \left(\frac{\partial y}{\partial v}\right)^{-2} \end{pmatrix} \quad (16)$$

and the material tensor is:

$$C = \frac{E}{1-v^2} \begin{pmatrix} a_{11}^2 & va_{11}a_{22} & 0 \\ va_{11}a_{12} & a_{22}^2 & 0 \\ 0 & 0 & \frac{1}{2}[(1-v)a_{11}a_{22}] \end{pmatrix} \quad (17)$$

The global curvatures for this plate are then given by:

$$\kappa_x = \frac{\partial^2 w}{\partial u^2} \cdot a_{11}; \quad \kappa_y = \frac{\partial^2 w}{\partial v^2} \cdot a_{22}; \quad \kappa_{xy} = 2 \cdot \frac{\partial^2 w}{\partial v \partial u} \cdot \sqrt{a_{11}a_{22}} \quad (18)$$

and the bending moments are:

$$m_x = D(\kappa_x + v \cdot \kappa_y) \quad (19)$$
$$m_y = D(\kappa_y + v \cdot \kappa_x) \quad (20)$$
$$m_{xy} = D(1-v)\kappa_{xy} \quad (21)$$

where $D = \frac{d^3 E}{12(1-v^2)}$ is the plate stiffness.

The components of the force vector due to a distributed load $q(x,y)$, normal to the plate mid surface are given by:

$$f_i = \int_{\xi=-1}^{1} \int_{\eta=-1}^{1} q(x,y) \cdot R_i^d(u,v) \cdot J \cdot du\, dv \quad (22)$$

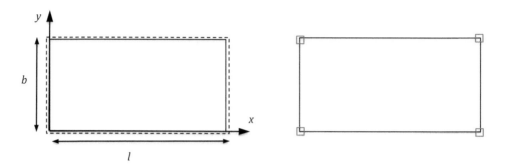

Figure 3 Geometry of plate and (right) representation with one NURBS patch ($p_u = p_v = 1$, $\Xi_u = \Xi_v = (0, 0, 1, 1)$) showing control points.

Dirichlet boundary conditions can be applied as explained previously by setting the parameters associated with the anchors located on the side to be restrained to zero (strong form) or adding a large number to the appropriate stiffness coefficients (weak form).

1.2 Examples

Rectangular plate The first example is a rectangular plate with the following properties:

- $E = 2.1 \cdot 10^8, \nu = 0$
- $d = 0.1, l = 20, b = 10$

The plate is simply supported on all sides and loaded by a vertical load of

$$q = q_0 \sin\left(\frac{\pi x}{l}\right) \sin\left(\frac{\pi y}{b}\right) \tag{23}$$

For this case we know the exact solution for the deflection:

$$w_{ex} = \frac{q_0}{\pi^4 D} \left(\frac{1}{l^2} + \frac{1}{b^2}\right)^{-2} \sin\left(\frac{\pi x}{l}\right) \sin\left(\frac{\pi y}{b}\right) \tag{24}$$

This geometry can be defined with a NURBS surface of order 1 in both u and v direction. Figure 3 shows the geometry and the representation with one NURBS patch. The simply supported boundary conditions are implemented by the strong form of the Dirichlet BC.

Order elevation Here we start with order elevating the NURBS describing the geometry, while leaving the geometry description untouched. Figure 4 shows the anchors for the various order elevations highlighting those where the parameters have been set to zero.

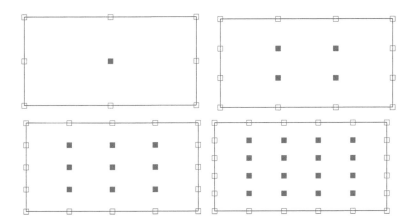

Figure 4 Location of anchors for various basis function orders $(p_u^d = p_v^d = 2$ to $5)$. Hollow squares depict anchors where the parameters have been set to zero to reflect zero Dirichlet boundary conditions.

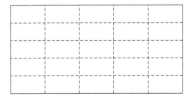

Figure 5 Subdivision of integration region for 4 knot insertions in each direction.

Knot insertion   This refinement strategy involves the repeated insertion of Knots after elevation to the minimum order of 2. Because of the limited span of the basis functions the plate has to be subdivided into integration regions as shown in Figure 5.

K-refinement   Here we elevate the order to $p = 3$ before inserting knots.
   To test the convergence we plot the error norm of the deflection defined as

$$\|\epsilon_w\| = \frac{\int_S |w - w_{ex}| dS}{\int_S |w_{ex}|} \tag{25}$$

where S is the surface of the plate, as a function of the degrees of freedom.
   In Figure 6 we show the convergence of the various refinement strategies. It is clear that the best refinement strategy for this type of problem is order elevation.

Skew plate   The plate is similar to the one presented previously except that it is no longer rectangular. We can obtain the new geometry by simply moving the lower right and left control point. However, to introduce the use of trimming used later for shells, we define the geometry with a trimmed NURBS patch. Figure 7 (left) shows the trimming curves and (right) the trimmed NURBS patch.

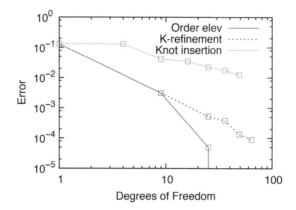

Figure 6 Plot of error versus degrees of freedom for various refinement strategies.

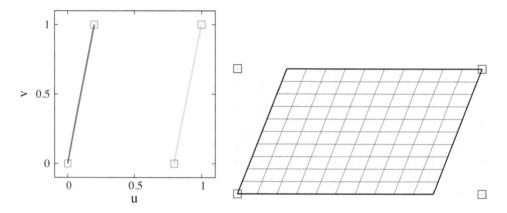

Figure 7 Trimming curves (left) and original and trimmed NURBS patch (right) for the skew plate.

In addition to the transformation between the u, v and s, t coordinate system already discussed in the section dealing with the description of the geometry, we also need to express the vectors v_1 and v_2 in the local s, t system. This means that we need the derivatives of x with respect to s, t.

These can be obtained applying the chain rule of differentiation:

$$v_1 = \frac{\partial x}{\partial t} = \frac{\partial x}{\partial u} \cdot \frac{\partial u}{\partial t} + \frac{\partial x}{\partial v} \cdot \frac{\partial v}{\partial t} \tag{26}$$

$$v_2 = \frac{\partial x}{\partial s} = \frac{\partial x}{\partial u} \cdot \frac{\partial u}{\partial s} + \frac{\partial x}{\partial v} \cdot \frac{\partial v}{\partial s}$$

For this problem no exact solution is available and therefore we plot the convergence of the maximum deflection in Figure 8. The results are compared to an untrimmed NURBS where the control points have been moved. We can see that the results are identical, thus verifying the trimming method. We apply the same refinement strategies as before and see that again order elevation gives the best results.

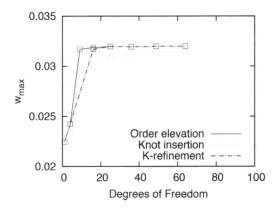

Figure 8 Convergence of maximum displacement for the different refinement strategies. Squares indicate the results for an untrimmed NURBS patch.

2 KIRCHHOFF SHELLS

For shells we have to include membrane effects, generalize the geometry to include curvature and introduce 3 components for the unknown $\mathbf{u}$ (u_x, u_y, u_z). The displacement field is now approximated by:

$$\mathbf{u} = \sum_{n=1}^{N_d} R_n^d(\mathbf{u}, \mathbf{v}) \mathbf{u}_n \tag{27}$$

The membrane strains are defined as

$$\epsilon_u = \sum_{n=1}^{N_d} \frac{\partial R_i^d}{\partial \mathbf{u}} \cdot \mathbf{v}_1 \cdot \mathbf{u}_n$$

$$\epsilon_v = \sum_{n=1}^{N_d} \frac{\partial R_i^d}{\partial \mathbf{v}} \cdot \mathbf{v}_2 \cdot \mathbf{u}_n \tag{28}$$

$$\gamma_{uv} = \sum_{n=1}^{N_d} \frac{1}{2} \left(\frac{\partial R_i^d}{\partial \mathbf{u}} \cdot \mathbf{v}_2 + \frac{\partial R_i^d}{\partial \mathbf{v}} \cdot \mathbf{v}_1 \right) \cdot \mathbf{u}_n$$

We now separate the membrane and the bending terms and compute two $\mathbf{B}$ matrices, $\mathbf{B}^m$ for the membrane terms and $\mathbf{B}^\kappa$ for the bending terms.

The matrix associated with the membrane strains is given by:

$$\mathbf{B}_i^m = \begin{pmatrix} m_1^i \cdot \mathbf{v}_{1x} & m_1^i \cdot \mathbf{v}_{1y} & m_1^i \cdot \mathbf{v}_{1z} \\ m_2^i \cdot \mathbf{v}_{2x} & m_2^i \cdot \mathbf{v}_{2y} & m_2^i \cdot \mathbf{v}_{2z} \\ 0.5(m_1^i \cdot \mathbf{v}_{2x} + m_2^i \cdot \mathbf{v}_{1x}) & 0.5(m_1^i \cdot \mathbf{v}_{2y} + m_2^i \cdot \mathbf{v}_{1y}) & 0.5(m_1^i \cdot \mathbf{v}_{2z} + m_2^i \cdot \mathbf{v}_{1z}) \end{pmatrix}$$
$$\tag{29}$$

where

$$m_1^i = \frac{\partial R_i^d}{\partial u}; \quad m_2^i = \frac{\partial R_i^d}{\partial v} \tag{30}$$

The matrix for the bending terms is given by:

$$\mathbf{B}_i^\kappa = \begin{pmatrix} b_{11}^i \cdot n_x & b_{11}^i \cdot n_y & b_{11}^i \cdot n_z \\ b_{22}^i \cdot n_x & b_{22}^i \cdot n_y & b_{22}^i \cdot n_z \\ b_{12}^i \cdot n_x & b_{12}^i \cdot n_y & b_{12}^i \cdot n_z \end{pmatrix} \tag{31}$$

where

$$b_{11}^i = \frac{\partial^2 R_i^d}{\partial u^2} - E_{11}^1 \cdot \frac{\partial R_i^d}{\partial u} - E_{11}^2 \cdot \frac{\partial R_i^d}{\partial v}$$

$$b_{22}^i = \frac{\partial^2 R_i^d}{\partial v^2} - E_{22}^1 \cdot \frac{\partial R_i^d}{\partial u} - E_{22}^2 \cdot \frac{\partial R_i^d}{\partial v} \tag{32}$$

$$b_{12}^i = 2 \cdot \left(\frac{\partial^2 R_i^d}{\partial u \partial v} - E_{12}^1 \cdot \frac{\partial R_i^d}{\partial u} - E_{12}^2 \cdot \frac{\partial R_i^d}{\partial v} \right)$$

and

$$E_{11}^1 = \frac{1}{J}\left[(\mathbf{v}_2 \times \mathbf{N}) \cdot \frac{\partial^2 \mathbf{x}}{\partial u^2} \right]; \quad E_{11}^2 = \frac{1}{J}\left[(\mathbf{N} \times \mathbf{v}_1) \cdot \frac{\partial^2 \mathbf{x}}{\partial u^2} \right]$$

$$E_{22}^1 = \frac{1}{J}\left[(\mathbf{v}_2 \times \mathbf{N}) \cdot \frac{\partial^2 \mathbf{x}}{\partial v^2} \right]; \quad E_{22}^2 = \frac{1}{J}\left[(\mathbf{N} \times \mathbf{v}_1) \cdot \frac{\partial^2 \mathbf{x}}{\partial v^2} \right] \tag{33}$$

$$E_{12}^2 = \frac{1}{J}\left[(\mathbf{v}_2 \times \mathbf{N}) \cdot \frac{\partial^2 \mathbf{x}}{\partial u \partial v} \right]; \quad E_{12}^2 = \frac{1}{J}\left[(\mathbf{N} \times \mathbf{v}_1) \cdot \frac{\partial^2 \mathbf{x}}{\partial u \partial v} \right]$$

The stiffness submatrix is now given by

$$\mathbf{k}_{ij} = \int_{\xi=-1}^{1} \int_{\eta=-1}^{1} \left[d \cdot (\mathbf{B}_i^m)^T \cdot \mathbf{C} \cdot \mathbf{B}_j^m + \frac{d^3}{12}(\mathbf{B}_i^\kappa)^T \cdot \mathbf{C} \cdot \mathbf{B}_j^\kappa \right] \cdot J \cdot 0.25 \cdot d\xi\, d\eta \tag{34}$$

2.1 Example 1: Scordelis roof

As the first example we choose one that has been used as a test case for shell analysis in many publications. It is known as the Scordelis-Lo[4] roof. The dimensions of the roof are shown in Figure 9.

The properties are:

- $E = 4.32 \cdot 10^8\, \text{kN/m}^2$, $v = 0$
- $d = 0.25$, $q = 90\, \text{kN/m}^2$

---

[4]Alexander Scordelis (1923–2007) was a structural engineer and a professor at the University of California, Berkeley. He made significant contributions to the analysis and design of long-span shell roofs, reinforced and prestressed concrete structures, and all types of bridges.

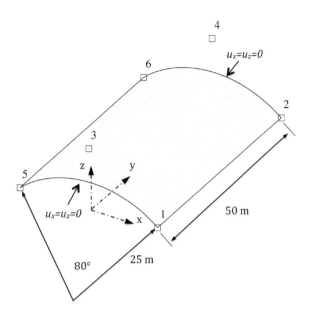

Figure 9 Geometry of shell and representation with one NURBS patch.

We describe the geometry with one NURBS patch. The input data for describing the geometry are:

```
Knot vectors:
2 1 0 0 1 1
3 2 0 0 0 1 1 1
Coefficients:
16.07 0 0 1
16.07 50 0 1
0 0 13.48 0.7661566
0 50 13.48 0.7661566
-16.07 0 0 1
-16.07 50 0 1
```

The geometry is shown in Figure 9. The roof is supported on each side as indicated. To avoid a rigid body movement one point was restrained in the y-direction. A reference solution is given in [4]. We obtain the solution in two ways:

- By order elevating the NURBS used for the description of the geometry
- Using B-splines with increasing orders

Figure 10 shows the convergence of the maximum displacement. We can see that order elevating the NURBS gives the best result. Using B-splines of different orders results in a slightly slower convergence.

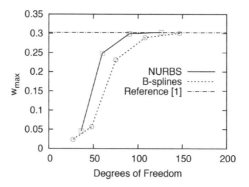

Figure 10 Convergence of solution for approximation with NURBS and B-splines.

Figure 11 Alexander Scordelis.

2.2 Example 2: Trimmed Scordelis roof

To apply the trimming method, used for the plate, to the shell we also need the second derivatives of $\mathbf{x}$ in terms of s, t. Using the chain rule of differentiation we obtain:

$$
\begin{aligned}
\frac{\partial^2 \mathbf{x}}{\partial s^2} &= \left(\frac{\partial^2 \mathbf{x}}{\partial u^2} \cdot \frac{\partial u}{\partial s} + \frac{\partial^2 \mathbf{x}}{\partial u \partial v} \cdot \frac{\partial v}{\partial s} \right) \cdot \frac{\partial u}{\partial s} + \frac{\partial \mathbf{x}}{\partial u} \cdot \frac{\partial^2 u}{\partial s^2} \\
&\quad + \left(\frac{\partial^2 \mathbf{x}}{\partial u \partial v} \cdot \frac{\partial u}{\partial s} + \frac{\partial^2 \mathbf{x}}{\partial v^2} \cdot \frac{\partial v}{\partial s} \right) \cdot \frac{\partial v}{\partial s} + \frac{\partial \mathbf{x}}{\partial v} \cdot \frac{\partial^2 v}{\partial s^2} \\
\frac{\partial^2 \mathbf{x}}{\partial t^2} &= \left(\frac{\partial^2 \mathbf{x}}{\partial u^2} \cdot \frac{\partial u}{\partial t} + \frac{\partial^2 \mathbf{x}}{\partial u \partial v} \cdot \frac{\partial v}{\partial t} \right) \cdot \frac{\partial u}{\partial t} + \frac{\partial \mathbf{x}}{\partial u} \cdot \frac{\partial^2 u}{\partial t^2} \\
&\quad + \left(\frac{\partial^2 \mathbf{x}}{\partial u \partial v} \cdot \frac{\partial u}{\partial t} + \frac{\partial^2 \mathbf{x}}{\partial v^2} \cdot \frac{\partial v}{\partial t} \right) \cdot \frac{\partial v}{\partial t} + \frac{\partial \mathbf{x}}{\partial v} \cdot \frac{\partial^2 v}{\partial t^2} \\
\frac{\partial^2 \mathbf{x}}{\partial t \partial s} &= \left(\frac{\partial^2 \mathbf{x}}{\partial u^2} \cdot \frac{\partial u}{\partial s} + \frac{\partial^2 \mathbf{x}}{\partial u \partial v} \cdot \frac{\partial v}{\partial s} \right) \cdot \frac{\partial u}{\partial t} + \frac{\partial \mathbf{x}}{\partial u} \cdot \frac{\partial^2 u}{\partial t \partial s} \\
&\quad + \left(\frac{\partial^2 \mathbf{x}}{\partial u^2} \cdot \frac{\partial u}{\partial s} + \frac{\partial^2 \mathbf{x}}{\partial v \partial u} \cdot \frac{\partial v}{\partial s} \right) \cdot \frac{\partial v}{\partial t} + \frac{\partial \mathbf{x}}{\partial v} \cdot \frac{\partial^2 v}{\partial t \partial s}
\end{aligned}
\tag{35}
$$

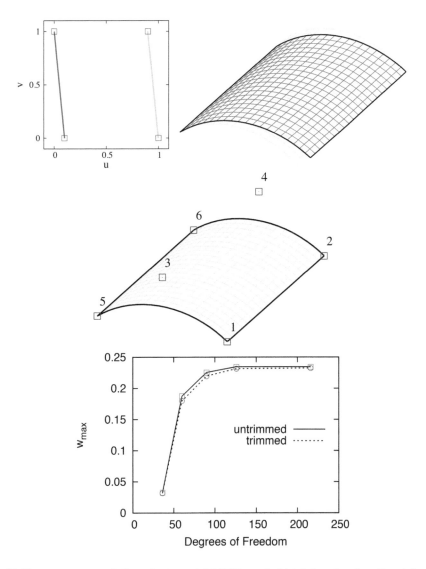

Figure 12 Trimming curves (left) and trimmed NURBS patch (right) for the skew Scordelis roof. Below: untrimmed NURBS patch with control points and convergence of maximum displacement.

With reference to Stage 2 the second derivatives of u, v are computed by:

$$\frac{\partial^2 u}{\partial s^2} = N_1(t) \cdot \frac{\partial^2 u_b(s)}{\partial s^2} + N_2(t) \cdot \frac{\partial^2 u_t(s)}{\partial s^2} \tag{36}$$

$$\frac{\partial^2 v}{\partial s^2} = N_1(t) \cdot \frac{\partial^2 v_b(s)}{\partial s^2} + N_2(t) \cdot \frac{\partial^2 v_t(s)}{\partial s^2} \tag{37}$$

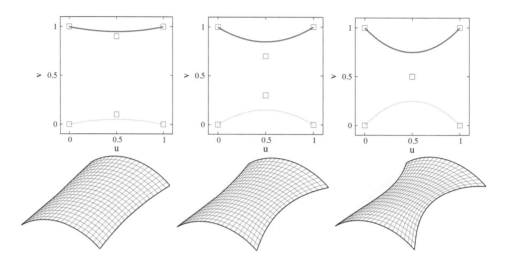

Figure 13 Arched Scordelis roof: Top: Trimming curves and bottom: trimmed surfaces. From left to right geometry 1 to 3.

$$\frac{\partial^2 u}{\partial t^2} = 0; \quad \frac{\partial^2 v}{\partial t^2} = 0 \tag{38}$$

$$\frac{\partial^2 u}{\partial s \partial t} = -1 \cdot \frac{\partial u_b}{\partial s} + 1 \cdot \frac{\partial u_t}{\partial s} \tag{39}$$

$$\frac{\partial^2 v}{\partial s \partial t} = -1 \cdot \frac{\partial v_b}{\partial s} + 1 \cdot \frac{\partial v_t}{\partial s} \tag{40}$$

For the first test we introduce skewness in the geometry by 2 linear trimming curves. Figure 12, shows the curves and geometry of the trimmed roof overlaid onto the original geometry. The same geometry can be defined without trimming by re-calculating the control point locations. The convergence of the maximum displacement, shown in Figure 12. It can be seen that the results for the trimmed and untrimmed shell are very close.

2.3   Example 3: Arched Scordelis roof

As a third example we introduce arching in the geometry, using trimming. The trimming curves and the trimmed surfaces are shown in Figure 13. The convergence of the maximum displacement using order elevation is shown in Figure 14. There is a decrease in maximum displacement for the first two geometries and then an increase. The reason for this is a change in the displacement pattern as can be seen in Figure 15.

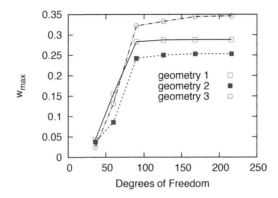

Figure 14 Arched Scordelis roof: Convergence of the maximum displacement.

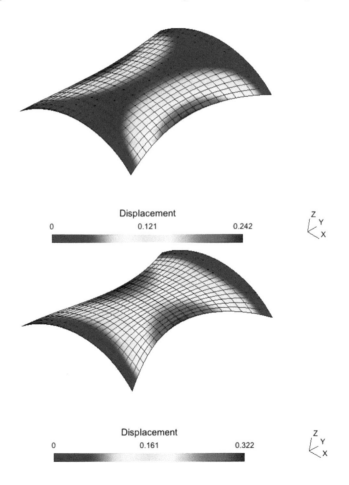

Figure 15 Contours of vertical displacement for geometries 2 and 3.

3 MULTIPLE PATCHES

To analyze a problem with multiple patches some changes have to be made to the program. Firstly, we must introduce a capability for handling more than one NURBS patch. Secondly, we have to ensure compatibility conditions are satisfied where the patches meet, i.e. we must enforce

$$\mathbf{u}_i^I = \mathbf{u}_i^{II} \tag{41}$$

where the subscript indicates the location on the interface and the superscript indicates the patch number. Note that unless we use special techniques such as proposed in [3] we are only able to enforce C^0 compatibility, i.e. the rotations will not match. It is no longer possible to use the order elevation of the NURBS describing the geometry. Therefore we will use B-splines and this means that we approximate the displacements by:

$$\mathbf{u} = \sum_{n=1}^{N_d} N_n^d(\mathrm{u}, \mathrm{v}) \mathbf{u}_n \tag{42}$$

where N_n^d are B-spline functions. Our proposed approach, that the approximation of the geometry is completely separate from the one of the unknown, is now crucial for the implementation.

To ensure C^0 compatibility at interfaces between patches the following conditions have to be satisfied:

- The same basis functions are used for the description of the unknown in the connecting patches at the interface.
- The location of anchors computed separately for each patch must match at the interface.

3.1 Assembly

The implementation of a simulation involving more than one NURBS patch starts with the determination of a *connectivity array* for each NURBS patch as introduced earlier. The connectivity array is created as follows:

- The locations of anchors are computed for each patch in the s, t coordinate system using the Greville formula and then mapped to the x, y, z coordinate system.
- It is checked whether any anchor locations coincide with ones computed for a previous patch.

  If this is the case, the anchor is given the number of the previously computed anchor.
  If not, the anchor is given a new number.

- Each NURBS patch is assigned a connectivity array that contains its anchor numbers.

The Dirichlet boundary conditions are implemented by producing a fixity array. This array is of size $(3, na)$ where na is the number of anchors and contains a 0 for each degree of freedom that is restrained and a 1 otherwise.

For the assembly, the connectivity vector is consolidated into a *destination vector* ldest of size $(3 * na)$. The entries in the destination vector specify the destination of each coefficient of the stiffness matrix of the patch in the global stiffness matrix. A zero entry specifies that the coefficient is not assembled because the associated parameter is known to be zero (due to an imposed zero boundary condition).

Part of the code to assemble the stiffness matrix is shown here:

```
for i=1:nas*3
  ii= ldest(i);
  if(ii == 0) continue endif
  F(ii,1)=F(ii,1) + f(i,1);
  for j=1:nas*3
    jj= ldest(j);
    if(jj == 0) continue endif
    K(ii,jj)=K(ii,jj) + k(i,j);
  end
 end
end
```

In the above *nas* is the number of anchors, F and K are the global force vector and stiffness matrix respectively. f and k is the force vector and stiffness matrix of the patch.

3.2 Example

We analyze a branched shell with two NURBS patches. Figure 16 shows the NURBS patches together with trimming curves and Figure 17 the trimmed geometry. The following properties are assumed

- $E = 10^7$
- $\nu = 0$
- $d = 0.2$

The shell is fixed as shown in Figure 18 and loaded by a vertical distributed load of 90 in the vertical direction. If we assume an order of $p = 2$ for the approximation in both directions and for both patches the connectivity arrays are given by:

```
Patch 1: 1    2    3    4    5    6    7    8    9
Patch 2: 7    10   11   8    12   13   9    14   15
```

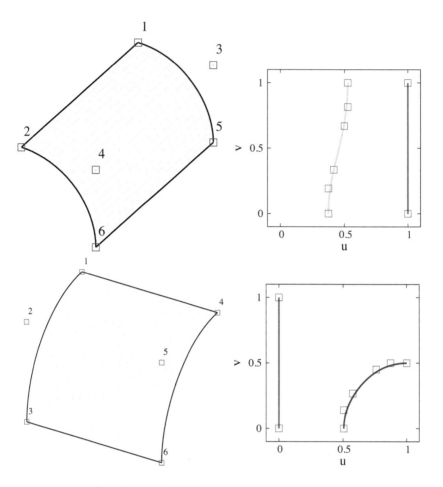

Figure 16 Two NURBS patches with trimming curves.

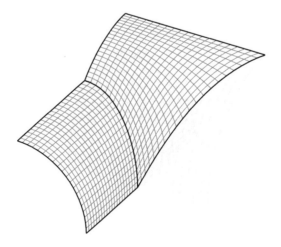

Figure 17 Branched shell with 2 trimmed NURBS patches.

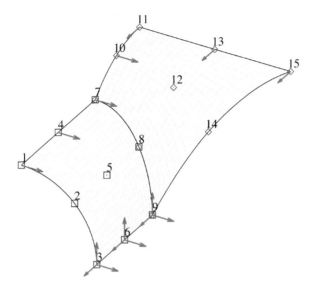

Figure 18 Location of anchors with global numbering for the approximation with B-splines of orders $p_u = 2, p_v = 2$ for both NURBS patches. Imposed boundary conditions are shown.

Figure 18 shows the location of the anchors and their numbering. It can be seen that the compatibility at the interface is ensured since the anchor locations match at the interface. For this example the fixity arrays are given by:

```
Patch 1:
    0    0    1    0    0    1    0    0    1
    1    0    1    1    0    1    1    0    1
    0    0    1    0    0    1    0    0    1
Patch 2:
    0    0    1    0    0    1    0    0    1
    1    1    0    0    0    0    0    0    0
    0    0    0    0    0    0    0    0    0
```

With this information the following destination vectors are obtained:

```
Patch 1:
1    0    2    3    4    5    0    0    0    6    0    7    8    9   10    0    0
0   11    0   12   13   14   15    0    0    0   47   48   49   50   51   52    0   53
54    0    0    0   55   56   57   58   59   60    0   61   62
Patch 2:
11    0   12   16    0   17    0   18   19   13   14   15   20   21   22    0   23   24
0    0    0   25   26   27    0   28   29   47   48   49   50   51   52    0   53
54    0    0    0   55   56   57   58   59   60    0   61   62
```

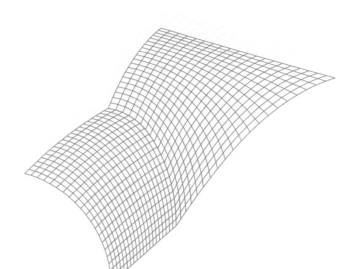

Figure 19 Results of the analysis: Displaced shape.

Figure 19 shows one result of the analysis namely the displaced shape.

4 SUMMARY AND CONCLUSIONS

In this stage we have ventured into three-dimensional space, but the geometry description only involved NURBS surfaces. This is because for some problems, such as thin plates and shells, simplified assumptions can be made that reduce the dimension of the problem by one, without affecting the quality of the results.

It should have become clear to the reader by this stage, that introducing NURBS technology has considerably improved the simulation of shells. Using conventional FEM the approximation of the unknown requires the consideration of rotational degrees of freedom and the generation of a mesh. Using NURBS surfaces for the description of the geometry and NURBS or B-splines for the approximation of the unknown the introduction of rotational degrees of freedom is not only avoided but also no mesh generation is required. NURBS and trimmed NURBS surfaces can describe complex geometries accurately, and therefore the first examples involved only one patch. However, it was shown how this can be extended to multiple patches for example in the case of intersecting shells.

Since the main theme of the book is to avoid mesh generation we have to look for alternatives to the formulations discussed so far, that are based on the Ritz method if we move to real 3-D problems. Fortunately such an alternative was proposed by Trefftz and this leads us to the integral equations discussed next.

BIBLIOGRAPHY

[1] Reinhard Fleissner. Isogeometrische Finite Elemente Methode für die lineare Kirchhoff-Love-Schale. Master's thesis, Technische Universitaet Graz, 2013.

[2] J. Kiendl, K.U. Bletzinger, J. Linhard, and R. Wüchner. Isogeometric shell analysis with Kirchhoff–Love elements. *Computer Methods in Applied Mechanics and Engineering*, 198(49):3902–3914, 2009.

[3] Josef M. Kiendl. *Isogeometric Analysis and Shape Optimal Design of Shell Structures*. PhD thesis, Technische Universitaet Muenchen, 2010.

[4] R. H. Macneal and R. L. Harder. A proposed standard set of problems to test finite element accuracy. *Finite Elements in Analysis and Design*, 1(1):3–20, 1985.

[5] J. N. Reddy. *Theory and analysis of elastic plates and shells*. CRC Press, Taylor and Francis, 2007.

Chapter 7

Stage 6: Integral equations

Sometimes one pays for the things one gets for nothing

A. Einstein

where we see the benefits of using integral equations.

I INTRODUCTION

So far we have followed the classical approach in modeling, i.e. we have approximated the unknown, using basis functions. We have seen that by using B-splines or NURBS we get an increase in accuracy with fewer parameters as compared with Lagrange polynomials. This is because these functions have properties that make them more suitable for approximation. Let us now look at an alternative to this approach. In choosing the functions for the approximation of the unknown we go one step further: We choose functions that exactly satisfy the differential equations. In elasticity this means that compatibility and equilibrium conditions are exactly satisfied, surely a big improvement. Therefore, we change the paradigm: We use functions that satisfy the DE exactly and approximate only the boundary conditions.

Functions that exactly satisfy the DE can be obtained as their solutions in an infinite domain. For potential problems this is a solution for a source concentrated at a point and in solid mechanics problems for point forces in an infinite domain. These solutions are also referred to as fundamental solutions or *Kernels*. The first idea of the method is due to Trefftz[1] ([10]) and was proposed as an alternative to the Ritz method introduced earlier.

Remark on notation   Here we use a mixture of indicial and matrix notation. In previous books on the BEM it was attempted to use only matrix notation but this was not successful and resulted in lengthy formulae. Unfortunately this means that two different notations are used for the coordinates i.e. x, y, z and x_1, x_2, x_3 or y_1, y_2, y_3. For the displacement components u_i are sometimes used instead of **u**. Note that u (italic) is used for the potential/temperature here, not to be confused with local coordinate u.

---

[1] German mathematician Erich Trefftz (1888 to 1937).

Figure 1 Erich Trefftz.

2 TREFFTZ METHOD

In the original version the method was applied to potential problems with the governing differential equation:

$$\frac{\partial q_i}{\partial x_i} - b = 0 \quad \text{for } i = 1, 2, (3)$$

(1)

where x_i are Cartesian coordinates, q_i are components of flow and b is flow per unit volume. This differential equation governs steady state flow of temperature, water and electricity.

For an isotropic material the relationship between potential or temperature u and flow is given by the *Fourier Law*:

$$q_i = -k \cdot \frac{\partial u}{\partial x_i}$$

(2)

where k is the permeability or conductivity. Substitution of (2) into (1) gives the *Laplace* equation

$$k \cdot \frac{\partial^2 u}{\partial x_i^2} - b = 0$$

(3)

A fundamental solution can be obtained for a source concentrated at a point in an unbounded (infinite) domain. For this we introduce the *Dirac Delta* function with the following properties:

$$\delta(\mathbf{y}, \mathbf{x}) = 0 \quad \text{for } \mathbf{y} \neq \mathbf{x}$$

(4)

$$\int_\Omega \delta(\mathbf{y}, \mathbf{x}) d\Omega = 1$$

where $\mathbf{y}$ (components y_1, y_2, y_3) denotes the location of the source (source point) and $\mathbf{x}$ (components x_1, x_2, x_3) any location (field point). The integral is over the infinite volume Ω.

A solution of the differential equation

$$k \cdot \frac{\partial^2 U}{\partial x_i^2} - \delta(\mathbf{y}, \mathbf{x}) = 0 \tag{5}$$

is, for a 3-D problem:

$$U(\mathbf{y}, \mathbf{x}) = \frac{1}{4\pi k} \frac{1}{r} \tag{6}$$

and for a 2-D problem:

$$U(\mathbf{y}, \mathbf{x}) = \frac{1}{2\pi k} \ln \frac{1}{r} \tag{7}$$

where capital U has been substituted for u to indicate that this is a fundamental solution and r is the distance between $\mathbf{x}$ and $\mathbf{y}$.

The flow vector at any point $\mathbf{x}$ can be obtained by substitution into Equation (2):

$$q_i(\mathbf{x}, \mathbf{y}) = -k \frac{\partial U(\mathbf{y}, \mathbf{x})}{\partial x_i} \tag{8}$$

The flow T in a direction defined by vector $\mathbf{n}$ can then be computed by

$$T(\mathbf{x}, \mathbf{y}) = -k \cdot \frac{\partial U}{\partial \mathbf{n}} = -k \cdot \left(n_x \frac{\partial U}{\partial x_1} + n_y \frac{\partial U}{\partial x_2} + n_z \frac{\partial U}{\partial x_3} \right) \tag{9}$$

For a 2-D problem we have

$$T(\mathbf{x}, \mathbf{y}) = \frac{1}{r} \frac{\cos \theta}{2\pi} \tag{10}$$

and for a 3-D problem

$$T(\mathbf{x}, \mathbf{y}) = \frac{1}{r^2} \frac{\cos \theta}{4\pi} \tag{11}$$

where

$$\cos \theta = n_x r_x + n_y r_y + n_z r_z \tag{12}$$

is the angle between the normalized position vector **r** and the vector **n** and

$$r_x = \frac{1}{r}(x_1 - y_1)$$

$$r_y = \frac{1}{r}(x_2 - y_2) \qquad\qquad (13)$$

$$r_z = \frac{1}{r}(x_3 - y_3)$$

It can be seen that both solutions are singular at **x** = **y**, i.e. tend to infinity as r goes to zero. The first one has a singularity of $O(\ln\frac{1}{r})$ or $O(\frac{1}{r})$ and is therefore *weakly singular*, the second has a singularity of $O(\frac{1}{r})$ or $O(\frac{1}{r^2})$ and is therefore *strongly singular*. This fact will pose some challenges in the implementation.

Functions for the fundamental solutions are shown below

```
function U= LapU((r,k,dim)
%----------------------------
%    Fundamental solution for Laplace
%    for Potential
%    Input:
%    r   ...   Distance between source and field point
%    k   ...   Conductivity
%    dim ...   Cartesian dimension (2-D,3-D)
%    Output:
%    U  ... Potential
%----------------------------
if(dim == 2)
    U= 1.0/(2.0*pi*k)*log(1/r);
else
    U= 1.0/(4.0*pi*r*k);
endif
endfunction;
```

```
function T= LapT(r,dxr,Vnorm,dim)
%----------------------------
%  Fundamental solution for Laplace
%  Normal gradient
%  Input:
%  r ...   Distance between source and field point
%  dxr(:) ...   rx , ry , rz
%  Vnorm(:)  ....   Normal vector
%  dim ...   Cartesian dimension
%  Output:
%  T  ...   flow in direction Vnorm
%----------------------------
```

```
if(dim == 2)
  T= vecdotp(Vnorm,dxr)/(2.0*pi*r);
else
  T= vecdotp(Vnorm,dxr)/(4.0*pi*r*r);
endif
endfunction;
```

The basic idea of the Trefftz method is to construct different fundamental solutions for a number of locations of fictitious source points with different source intensities. For a source intensity F_1 at location $\mathbf{y}_1$ the resulting flow is for example $T(\mathbf{y}_1, \mathbf{x}) \cdot F_1$. Assume that we have a boundary value problem, where the flow $t(\mathbf{x})$ in a direction normal to the boundary (defined by $\mathbf{n}$) is given and the solution for the potential u is required. We can satisfy the boundary conditions point wise (i.e. at $\mathbf{x}_1$, $\mathbf{x}_2$, ...) by assuming they are composed of fundamental solutions times unknown source densities at a number of source points N.

For example for the boundary condition at $\mathbf{x}_1$ we have

$$t(\mathbf{x}_1) = \sum_{n=1}^{N} T(\mathbf{y}_n, \mathbf{x}_1) \cdot F_n \tag{14}$$

We can now satisfy the boundary conditions at N points and get a square system of equations

$$\mathbf{t} = \mathbf{T} \cdot \mathbf{F} \tag{15}$$

where

$$\mathbf{T} = \begin{pmatrix} T(\mathbf{y}_1, \mathbf{x}_1) & T(\mathbf{y}_2, \mathbf{x}_1) & \cdots \\ T(\mathbf{y}_1, \mathbf{x}_2) & T(\mathbf{y}_2, \mathbf{x}_2) & \cdots \\ \cdots & \cdots & \cdots \end{pmatrix} \tag{16}$$

and

$$\mathbf{t} = \begin{pmatrix} t(\mathbf{x}_1) \\ t(\mathbf{x}_2) \\ \vdots \end{pmatrix}, \quad \mathbf{F} = \begin{pmatrix} F_1 \\ F_2 \\ \vdots \end{pmatrix} \tag{17}$$

which can be solved for the *fictitious* source intensities. We use the term *fictitious* to indicate that the sources are not really there, are only used to obtain a solution of the problem and must always be placed outside the solution domain.

After obtaining the solution for $\mathbf{F}$ the results for u at any point $\mathbf{x}$ inside the solution domain can be computed by

$$u(\mathbf{x}) = \sum_{n=1}^{N} U(\mathbf{y}_n, \mathbf{x}) \cdot F_n \tag{18}$$

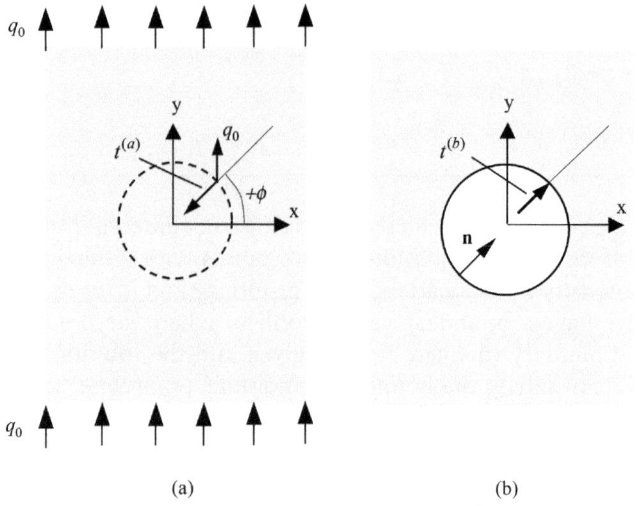

(a)   (b)

Figure 2   Example for the Trefftz method.

Similarly the flow vector can be obtained as

$$q_i(\mathbf{x}) = -k \cdot \sum_{n=1}^{N} \frac{\partial U(\mathbf{y}_n, \mathbf{x})}{\partial x_i} \cdot F_n \tag{19}$$

2.1   Example

As an example we show the plane flow past a circular insulator. Consider the problem where there is a constant flow q_0 in the y-direction. We analyze the change in flow as we place an insulator with radius R into the flow, which means that no flow can occur perpendicular to the boundary of the insulator. The problem can be split into two (see Figure 2):

- Problem a: flow without insulator
- Problem b: change in flow due to insulator

Problem (a) has an exact solution:

$$u^a(\mathbf{x}) = -\frac{q_0}{k} \cdot y \tag{20}$$

To obtain the boundary conditions for problem (b) we compute the flow normal to the boundary of the insulator:

$$t^a = n_y \cdot q_0 = -q_0 \cdot \sin\phi \tag{21}$$

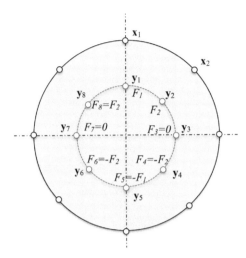

Figure 3 Figure showing the location of field and source points as well as the source intensities *F*.

Table 1 Coefficients of **T**.

i	$T(\mathbf{y}_i, \mathbf{x}_1)$	$T(\mathbf{y}_i, \mathbf{x}_2)$	multiplied with
1	0.53052	0.16073	F_1
2	0.16073	0.53052	F_2
4	0.09594	0.10681	$-F_2$
5	0.09362	0.09594	$-F_1$
6	0.09594	0.09362	$-F_2$
8	0.16073	0.10681	F_2

Since the final flow has to be zero, the boundary condition for problem b is $t^b = -t^a$. We now solve the problem (b) with the Trefftz method. For this we specify a number of points $\mathbf{x}_j$ where the boundary condition will be satisfied exactly. This is matched by source points $\mathbf{y}_j$ which are placed outside the solution space, i.e. inside the insulator along a circle with the radius cR where $0 < c < 1$ must be chosen so that a singularity is avoided and the points are not too close to each other.

The most optimal value of c was found to be 0.7. We apply Equation (15) for the solution and exploit the symmetries of the problem. As shown in Figure 3 only two unknown fictitious sources (F_1 and F_2) remain. Therefore we need only two equations, i.e. we need to satisfy the boundary conditions only at points $\mathbf{x}_1$ and $\mathbf{x}_2$. Because of the symmetry conditions the other boundary conditions are also satisfied. Table 1 shows the coefficients of the matrix **T** and the source intensities they are multiplied with.

We end up with a reduced matrix **T**:

$$\mathbf{T} = \begin{pmatrix} 0.43690 & 0.12958 \\ 0.06479 & 0.43690 \end{pmatrix} \tag{22}$$

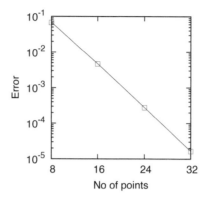

Figure 4 Convergence of maximum potential.

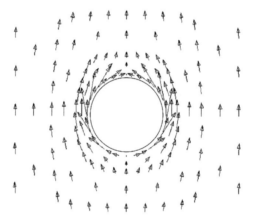

Figure 5 Results of Trefftz method: flow vectors.

and a reduced right hand side:

$$\mathbf{F} = \begin{pmatrix} 1.0 \\ 0.7071 \end{pmatrix} \tag{23}$$

which can be solved for the fictitious sources F_1 and F_2. We use Equations (18) for computing the potential and (19) for the flow vectors.

The exact solution for this problem is known. The accuracy of the solution depends of the number of points $\mathbf{x}$ but convergence for this simple problem is quite fast. Figure 4 shows the convergence of the maximum potential as a plot of error versus the total number of boundary points (not considering symmetry). The resulting flow vectors are plotted in Figure 5. More details can be found in [2] and [3].

2.2 Conclusions

The Trefftz method is very simple but is not applicable to practical problems for the following reasons:

- It is not very user friendly to have to choose the locations of the fictitious sources and for practical problems this is not easy to do.
- The location where fictitious sources are placed can not be chosen freely: They must lie outside the computational domain, they must not be too close together and not too near to the boundary.
- There is no control of what happens in between the points where the boundary conditions are enforced, leaving the possibility of oscillatory behavior.

Obviously the solution is to reduce the error in satisfying the boundary condition over the whole boundary, rather than pointwise. This leads us to integral equations.

3 INTEGRAL EQUATIONS

We explain how to obtain an integral statement on a problem in elasticity first, since the derivation uses well known engineering principles. The governing differential equation is

$$\frac{\partial \sigma_{jk}}{\partial x_k} + b_j = 0 \quad i = 1, 2, (3), \quad j = 1, 2, (3) \tag{24}$$

where σ_{jk} is the stress tensor and b_j is the body force vector.

Introducing the constitutive law

$$\begin{aligned} \sigma_{ij} &= \lambda \delta_{ij} \varepsilon_{kk} + 2\mu \varepsilon_{ij} \\ &= \lambda \delta_{ij} u_{k,k} + \mu (u_{i,j} + u_{j,i}) \end{aligned} \tag{25}$$

where the Lame constants are given by:

$$\lambda = \frac{Ev}{(1+v)(1-2v)}, \quad \mu = \frac{E}{2(1+v)} \tag{26}$$

E is the Modulus of Elasticity and v is the Poisson's ratio. In the above ϵ_{ij} is the strain tensor, u_i is the displacement vector and $u_{i,j}$ means the first derivative to j. After substitution we obtain the differential equation in terms of displacements u_i

$$\lambda u_{k,ki} + \mu (u_{i,jj} + u_{j,ij}) + b_i = 0 \tag{27}$$

where $u_{k,ki}$ means the second derivative to k, i.

A fundamental solution of this DE for a plane problem can be found by introducing the *Dirac Delta* function for the body force b_i i.e. we obtain 2 solutions for the displacement vector at $\mathbf{x}$: for forces $b_1 = \delta(\mathbf{y}, \mathbf{x})$, $b_2 = 0$ and $b_1 = 0$, $b_2 = \delta(\mathbf{y}, \mathbf{x})$.

Changing to matrix notation, the fundamental solutions for the displacement is given by

$$\mathbf{U}(\mathbf{y}, \mathbf{x}) = \begin{pmatrix} U_{xx} & U_{xy} \\ U_{yx} & U_{yy} \end{pmatrix} = C \ln\left(\frac{1}{r}\right) (C_1 \mathbf{I} + \mathbf{R}) \tag{28}$$

where the first subscript in U defines the direction of the point force and the second the direction of the displacement. For plane strain, the constants are given by $C = \frac{1}{8\pi\mu(1-\nu)}$ and $C_1 = 3 - 4\nu$. $\mathbf{I}$ is the unit matrix and

$$\mathbf{R} = \begin{pmatrix} r_x^2 & r_x r_y \\ r_y r_x & r_y^2 \end{pmatrix} \tag{29}$$

As expected $\mathbf{U}$ is symmetric.

From this, a fundamental solution for the stresses can be derived. The stresses acting on a plane perpendicular to a vector $\mathbf{n}$, which will be called *tractions* from now on, are obtained by

$$\mathbf{T}(\mathbf{y}, \mathbf{x}) = C_2 \frac{1}{r}((C_3 \mathbf{I} + 2\,\mathbf{R}) \cos\theta + C_3 \mathbf{R}_1) \tag{30}$$

where for plane strain $C_2 = \frac{1}{4\pi(1-\nu)}$, $C_3 = 1 - 2\nu$ and

$$\mathbf{R}_1 = \begin{pmatrix} 0 & n_x r_y - n_y r_x \\ n_y r_x - n_x r_y & 0 \end{pmatrix} \tag{31}$$

It can be seen that $\mathbf{T}$ consists of a symmetric part multiplied by $\cos\theta$ and an anti-symmetric part represented by $\mathbf{R}_1$. Listings of functions to compute the fundamental solutions are shown below:

```
function UK   = UKernel(xP,xQ,E,ny)
% ---------------------------------------
% Computes U(y,x) for plane strain
% INPUT:
% xP ...   Coordinates of source point
% xQ ...   Coordinates of field point
% E  ... modulus of elasticity
% ny ... Poissons ratio
%
% OUTPUT:
% UK   ... Displacement Kernel
%---------------------------------------
r= ((xQ(1)-xP(1))^2+(xQ(2)-xP(2))^2)^0.5;
dxr(1)= (xQ(1)-xP(1))/r; dxr(2)= (xQ(2)-xP(2))/r;
G= E/(2.0*(1+ny)); c1= 3.0 - 4.0*ny; c= 1.0/(8.0*pi*G*(1.0 - ny));
clog= -c1*log(r);
UK(1,1)= c*(clog + dxr(1)*dxr(1)); UK(1,2)= c*dxr(1)*dxr(2);
UK(2,2)= c*(clog + dxr(2)*dxr(2)); UK(2,1)= UK(1,2);
endfunction
```

```
function TK  = TKernel(xP,xQ,Vnor,ny)
%---------------------------------------------------------
% TKernel: Computes T(y,x) for plane strain
%
% INPUT:
% xP ...    Coordinates of source point
% xQ ...    Coordinates of field point
% Vnor ... vector normal to boundary
% ny  ... Poissons ratio
%
% OUTPUT:
% TK   ... Traction Kernel
%---------------------------------------------------------
r= ((xQ(1)-xP(1))^2+(xQ(2)-xP(2))^2)^0.5;
dxr(1)= (xQ(1)-xP(1))/r; dxr(2)= (xQ(2)-xP(2))/r;
c3= 1.0 - 2.0*ny; Costh= vecdot(Vnor,dxr);
c2= 1.0/(4.0*pi*(1.0 - ny)); Conr= c2/r;
TK(1,1)=-(Conr*(c3 + 2.0*dxr(1)*dxr(1))*Costh);
TK(1,2)=-(Conr*(2.0*dxr(1)*dxr(2)*Costh+c3*(Vnor(1)*dxr(2)-Vnor(2)*dxr(1))));
TK(2,2)=-(Conr*(c3 + 2.0*dxr(2)*dxr(2))*Costh);
TK(2,1)=-(Conr*(2.0*dxr(1)*dxr(2)*Costh+c3*(Vnor(2)*dxr(1)-Vnor(1)*dxr(2))));
endfunction
```

For 3-D problems the fundamental solutions are

$$\mathbf{U(y,x)} = C\frac{1}{r}(C_1\mathbf{I} + \mathbf{R}) \tag{32}$$

where $C = \frac{1}{16\pi\mu(1-\nu)}$ and $C_1 = 3 - 4\nu$, r is the distance between point $\mathbf{y}$ and $\mathbf{x}$ and

$$\mathbf{R} = \begin{pmatrix} r_x^2 & r_xr_y & r_xr_z \\ r_yr_x & r_y^2 & r_yr_z \\ r_zr_x & r_zr_y & r_z^2 \end{pmatrix} \tag{33}$$

Below we show the listing of the function to compute the U-Kernel

```
function UK  = UKernel(r,dxr,E,ny)
%-------------------------------
% Computes U(y,x) for 3-D
%
% INPUT:
% r ...    distance field point source point
% dxr ...    rx,ry,rz
% E  ... modulus of elasticity
```

```
% ny ... Poissons ratio
%
% OUTPUT:
% UK   ... Displacement Kernel
%-------------------------------
G= E/(2.0*(1+ny)); c1= 3.0 - 4.0*ny;
c= 1.0/(16.0*pi*G*(1.0 - ny));conr=c/r;
UK(1,1)= conr*(c1 + dxr(1)*dxr(1));
UK(1,2)= conr*dxr(1)*dxr(2); UK(1,3)= conr*dxr(1)*dxr(3);
UK(2,1)= UK(1,2); UK(2,2)= conr*(c1 + dxr(2)*dxr(2));
UK(2,3)= conr*dxr(2)*dxr(3); UK(3,1)= UK(1,3);UK(3,2)= UK(2,3);
UK(3,3)= conr*(c1 + dxr(3)*dxr(3));
endfunction;
```

The fundamental solution for the tractions is given by

$$\mathbf{T}(P,Q) = C_2 \frac{1}{r^2}((C_3\mathbf{I} + 3\,\mathbf{R}) * \cos\theta + C_3\mathbf{R}_1) \tag{34}$$

where $C_2 = \frac{1}{8\pi(1-\nu)}$, $C_3 = 1 - 2\nu$ and

$$\mathbf{R}_1 = \begin{pmatrix} 0 & n_x r_y - n_y r_x & n_x r_z - n_z r_x \\ n_y r_x - n_x r_y & 0 & n_y r_z - n_z r_y \\ n_z r_x - n_x r_z & n_z r_y - n_y r_z & 0 \end{pmatrix} \tag{35}$$

$\mathbf{T}$ can be split into a symmetric and antisymmetric part.

$$\mathbf{T} = \frac{1}{r^2}(\mathbf{T}_s + \mathbf{T}_a) \tag{36}$$

Figure 6 Enrico Betti.

where

$$\mathbf{T}^s = C_2(C_3\mathbf{I} + 3\mathbf{R}) \cdot \cos\theta$$

$$\mathbf{T}^a = C_3\mathbf{R}_1$$

(37)

```
function TK   = TKernel(r,dxr,Vnor,ny)
%-------------------------------------------------
% TKernel: Computes T(P,Q) for 3-D
%
% INPUT:
% r ...    distance source point field point
% dxr ...    rx,ry,rz
% Vnor ... vector normal to boundary
% ny  ... Poissons ratio
%
% OUTPUT:
% TK   ... Traction Kernel
%-------------------------------------------------
c3= 1.0 - 2.0*ny; Costh= vecdotp(Vnor,dxr);
c2= 1.0/(8.0*pi*(1.0 - ny)); Conr= c2/r^2;
TK(1,1)= -Conr*(c3 + 3.0*dxr(1)*dxr(1))*Costh;
TK(1,2)= -Conr*(3.0*dxr(1)*dxr(2)*Costh-c3*(Vnor(2)*dxr(1)-Vnor(1)*dxr(2)));
TK(1,3)= -Conr*(3.0*dxr(1)*dxr(3)*Costh-c3*(Vnor(3)*dxr(1)-Vnor(1)*dxr(3)));
TK(2,1)= -Conr*(3.0*dxr(1)*dxr(2)*Costh-c3*(Vnor(1)*dxr(2)-Vnor(2)*dxr(1)));
TK(2,2)= -Conr*(c3 + 3.0*dxr(2)*dxr(2))*Costh;
TK(2,3)= -Conr*(3.0*dxr(2)*dxr(3)*Costh-c3*(Vnor(3)*dxr(2)-Vnor(2)*dxr(3)));
TK(3,1)= -Conr*(3.0*dxr(1)*dxr(3)*Costh-c3*(Vnor(1)*dxr(3)-Vnor(3)*dxr(1)));
TK(3,2)= -Conr*(3.0*dxr(2)*dxr(3)*Costh-c3*(Vnor(2)*dxr(3)-Vnor(3)*dxr(2)));
TK(3,3)= -Conr*(c3 + 3.0*dxr(3)*dxr(3))*Costh;
endfunction;
```

3.1 Theorem of Betti

To obtain an integral statement we apply the well known theorem of Betti[2]. We specify the first load case to be the one with a fictitious source (unit point loads in x and y directions) placed at $\mathbf{y}$ (displacements $\mathbf{U}(\mathbf{y},\mathbf{x})$, tractions $\mathbf{T}(\mathbf{y},\mathbf{x})$) and the second load case to be the real load case that we want to solve (displacements $\mathbf{u}(\mathbf{x})$, tractions $\mathbf{t}(\mathbf{x})$). We assume for the moment that no body forces (other than the *Dirac Delta* forces for the first load case) are present in the domain. These will be introduced later.

---

[2]Betti's theorem, also known as Maxwell-Betti reciprocal work theorem, discovered by Enrico Betti in 1872, states that for a linear elastic structure subject to two sets of forces P_i $i = 1, \ldots, m$ and Q_j, $j = 1, 2, \ldots, n$, the work done by the set P through the displacements produced by the set Q is equal to the work done by the set Q through the displacements produced by the set P.

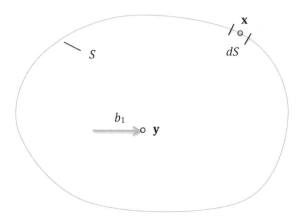

Figure 7 Explanation of the application of Betti's theorem.

We consider the work done by the tractions times the displacements along a contour S and the work done by the *Dirac Delta* forces times the displacements (see Figure 7).

The work done by the displacements of load case 1 times the tractions of load case 2 on a small portion of S, dS is:

$$dW_{12} = U(y, x) \cdot t(x) \cdot dS \tag{38}$$

The work done by the tractions of load case 1 times the displacements of load case 2 is:

$$dW_{21} = T(y, x) \cdot dS \cdot u(x) \tag{39}$$

The work done by displacements of load case 1 times the loads of load case 2 at point **y** is:

$$W_{12}(y) = 0 \tag{40}$$

since no body forces are present.

The work done by displacements of load case 2 times the loads of load case 1 is:

$$W_{21}(y) = I \cdot u(y) \tag{41}$$

Integrating over S and setting the $W_{12} = W_{21}$ we obtain

$$I \cdot u(y) = \int_S U(y, x) t(x) dS - \int_S T(y, x) u(x) dS \tag{42}$$

Figure 8 Carlo Somigliana.

This integral equation is also known as the Somigliana identity[3]. The Somigliana integral equation for elasticity is the equivalent of Green's formula for potential theory.

For potential problems we apply the Green's formula (for a derivation see [3]) and obtain

$$u(\mathbf{y}) = \int_S U(\mathbf{y}, \mathbf{x})t(\mathbf{x})dS - \int_S T(\mathbf{y}, \mathbf{x})u(\mathbf{x})dS \tag{43}$$

It is inconvenient to have points **y** inside the domain and points **x** on the boundary. Ideally we would like to have all the points on the boundary. Moving points **y** to the boundary involves a limiting process of the integral as **y** approaches **x**.

Consider the 2-D case in Figure 9 where we provide a circular region of exclusion around the point **y**, S_ϵ.

We can rewrite the integrals as

$$\int_S \mathbf{U}(\mathbf{y}, \mathbf{x})\mathbf{t}(\mathbf{x})dS = \int_{S_\epsilon} \mathbf{U}(\mathbf{y}, \mathbf{x})\mathbf{t}(\mathbf{x})dS + \int_{S-S\epsilon} \mathbf{U}(\mathbf{y}, \mathbf{x})\mathbf{t}(\mathbf{x})dS$$

$$\int_S \mathbf{T}(\mathbf{y}, \mathbf{x})\mathbf{t}(\mathbf{x})dS = \int_{S_\epsilon} \mathbf{T}(\mathbf{y}, \mathbf{x})\mathbf{t}(\mathbf{x})dS + \int_{S-S\epsilon} \mathbf{T}(\mathbf{y}, \mathbf{x})\mathbf{t}(\mathbf{x})dS \tag{44}$$

---

[3]Carlo Somigliana (1860 to 1955) was an Italian mathematician and a classical mathematical physicist faithful to the school of Enrico Betti and Eugenio Beltrami. He made important contributions in elasticity.

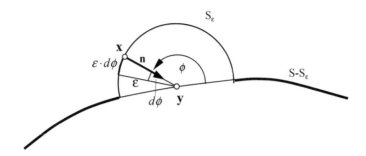

Figure 9 Figure explaining limiting process.

We examine the value of the integral over S_ϵ as the radius of the arc ϵ tends to zero, $\mathbf{t}(\mathbf{x})$ tends to $\mathbf{t}(\mathbf{y})$ and becomes independent of the integral. Applying cylindrical coordinates we have:

$$\lim_{\epsilon \to 0} \int_{S_\epsilon} \mathbf{U}(\mathbf{y},\mathbf{x})\mathbf{t}(\mathbf{x})dS = \mathbf{t}(\mathbf{y}) \int_0^\pi \mathbf{U} \cdot \epsilon \cdot d\phi = 0 \tag{45}$$

For the symmetric part of the Kernel $\mathbf{T}$ we have:

$$\lim_{\epsilon \to 0} \int_{S_\epsilon} \mathbf{T}^s(\mathbf{y},\mathbf{x})\mathbf{u}(\mathbf{x})dS = \mathbf{u}(\mathbf{y}) \int_0^\pi \frac{\mathbf{C} \cdot \cos\theta}{\epsilon} \cdot \epsilon \cdot d\phi \tag{46}$$

and with $\cos\theta = -1$

$$\mathbf{u}(\mathbf{y}) \int_0^\pi \frac{-1}{2\pi} \cdot d\phi = -\frac{1}{2} \cdot \mathbf{u}(\mathbf{y}) \tag{47}$$

The result for the antisymmetric part of the Kernel is:

$$\lim_{\epsilon \to 0} \int_{S_\epsilon} \mathbf{T}^a(\mathbf{y},\mathbf{x})\mathbf{u}(\mathbf{x})dS = 0 \tag{48}$$

The integral equation can now be written as:

$$\mathbf{c} \cdot \mathbf{u}(\mathbf{y}) = \lim_{\epsilon \to 0} \left[\int_{S-S_\epsilon} \mathbf{U}(\mathbf{y},\mathbf{x}) \cdot \mathbf{t}(\mathbf{x})dS - \int_{S-S_\epsilon} \mathbf{T}(\mathbf{y},\mathbf{x}) \cdot \mathbf{u}(\mathbf{x})dS \right] \tag{49}$$

Where for smooth boundaries the *free term* is given by

$$\mathbf{c} = \frac{1}{2}\mathbf{I} \tag{50}$$

Similarly the integral equation for potential problems is obtained as

$$c \cdot u(\mathbf{y}) = \lim_{\epsilon \to 0} \left[\int_{S-S_\epsilon} U(\mathbf{y}, \mathbf{x}) \cdot t(\mathbf{x}) dS - \int_{S-S_\epsilon} T(\mathbf{y}, \mathbf{x}) \cdot u(\mathbf{x}) dS \right] \tag{51}$$

where $c = \frac{1}{2}$. For boundaries that are not smooth the value of c differs.

However, as we will see in a moment this term is never explicitly computed. Because of the singularity of the Kernels the integrands approach infinity as $\mathbf{x}$ approaches $\mathbf{y}$ and care has to be taken in evaluating the integrals. Indeed, the second integral only exists as a Cauchy principal value[4]. It should be noted that the integral equations are equally valid for a finite domain (where the domain lies inside the boundary contour) and an infinite domain (where the domain lies outside). Indeed the definition of the problem would be incomplete, if only the boundary is specified. In addition a specification (via the outward normal) of the direction away from the material is required.

3.2 Rigid body trick

A clever way of avoiding the computation of the *free term* and the Cauchy principal value is to apply a rigid body movement of a finite domain (i.e. we set the displacement values at all points on the boundary equal). In this case the boundary tractions must be zero. Assuming a constant displacement vector $\mathbf{u}_r$ at all locations on the boundary ($\mathbf{x}$ and $\mathbf{y}$) and setting all tractions on the boundary, $\mathbf{t}(\mathbf{x})$, to zero we obtain the following integral equations:

$$\mathbf{c} \cdot \mathbf{u}_r = -\int_S \mathbf{T}(\mathbf{y}, \mathbf{x}) dS \cdot \mathbf{u}_r \tag{52}$$

or

$$\mathbf{c} = -\int_S \mathbf{T}(\mathbf{y}, \mathbf{x}) dS \tag{53}$$

Substitution of this result into (49) yields

$$-\int_S \mathbf{T}(\mathbf{y}, \mathbf{x}) dS \cdot \mathbf{u}(\mathbf{y}) = \int_S \mathbf{U}(\mathbf{y}, \mathbf{x}) \cdot \mathbf{t}(\mathbf{x}) dS - \int_S \mathbf{T}(\mathbf{y}, \mathbf{x}) \cdot \mathbf{u}(\mathbf{x}) dS \tag{54}$$

or

$$\int_S \mathbf{U}(\mathbf{y}, \mathbf{x}) \cdot \mathbf{t}(\mathbf{x}) dS = \int_S \mathbf{T}(\mathbf{y}, \mathbf{x}) \cdot (\mathbf{u}(\mathbf{x}) - \mathbf{u}(\mathbf{y})) dS \tag{55}$$

We have not only succeeded in getting rid of the free term but also in removing the need to compute a Cauchy principal value for the second integral because when $\mathbf{x}$ approaches $\mathbf{y}$ the Kernel is multiplied by a term approaching zero.

---

[4]The Cauchy principal value, named after Augustin Louis Cauchy, is a method for assigning values to certain improper integrals which would otherwise be undefined.

With modification, the trick can also be applied to problems involving an infinite domain, where constant displacement values along the boundary would not result in zero tractions. In this case we make the domain finite by introducing an artificial circular or spherical boundary with radius R and letting R go to infinity. This means that when we apply the rigid body motion we have to include the integral over the artificial boundary S_p (also known as the *azimuthal integral*).

$$\mathbf{c} = -\left(\int_S \mathbf{T}(\mathbf{y}, \mathbf{x}) dS + \int_{S_p} \mathbf{T}(\mathbf{y}, \mathbf{x}) dS_p \right) \tag{56}$$

The azimuthal integral can be evaluated analytically using polar coordinates:

$$\mathbf{A} = \int_{S_p} \mathbf{T}(\mathbf{y}, \mathbf{x}) dS_p = \int_0^{2\pi} \frac{1}{R} (\mathbf{T}^s + \mathbf{T}^a) R \, d\phi = -\mathbf{I} \tag{57}$$

Since R cancels out there is no need to take the limit $R \to \infty$ and this result is also valid for an infinite domain. For a semi-infinite domain the limits of the integral are from 0 to π and the result is $\mathbf{A} = \frac{1}{2}\mathbf{I}$.

The resulting integral equations for an infinite domain are

$$\int_S \mathbf{U}(\mathbf{y}, \mathbf{x}) \cdot \mathbf{t}(\mathbf{x}) dS = \int_S \mathbf{T}(\mathbf{y}, \mathbf{x}) \cdot (\mathbf{u}(\mathbf{x}) - \mathbf{u}(\mathbf{y})) dS - \mathbf{A} \tag{58}$$

A similar trick can be applied to potential problems: For a uniform temperature the flow must be zero, resulting in:

$$\int_S U(\mathbf{y}, \mathbf{x}) \cdot t(\mathbf{x}) dS = \int_S T(\mathbf{y}, \mathbf{x}) \cdot (u(\mathbf{x}) - u(\mathbf{y})) dS \tag{59}$$

for finite problems and

$$\int_S U(\mathbf{y}, \mathbf{x}) \cdot t(\mathbf{x}) dS = \int_S T(\mathbf{y}, \mathbf{x}) \cdot (u(\mathbf{x}) - u(\mathbf{y})) dS - A \tag{60}$$

for infinite problems.

Exterior and interior domains   The unique feature of the boundary integral representation is that one needs to specify the side of the boundary where the material is and if it is a finite domain (interior) or an infinite or semi-infinite (exterior) domain. The first definition is via a vector normal to the boundary that points away from the material (*outward normal*) and the second one is by adding or not the azimuthal integral.

3.3   Conclusions

We have now succeeded to replace the differential equation with a boundary integral equation. Historically this was a remarkable achievement. In 1860 Du Bois-Reymond[5] wrote (translation from German):

> "Since 1852 I have come across integral equations in the theory of partial differential equations so often that I am convinced that advances in this theory are linked to the treatment of integral equations, about which virtually nothing is known."

In 1930 E.J. Nyström [6] wrote (translated from German):

> "Indeed the later development in science has not only verified his statement, but integral equations and their applications have developed into a powerful tool that may be even more general and more useful than that of differential equations. For example recently an integral equation was developed for which there is no known differential equation"

This statement comes about 25 years before the advent of the digital computer. Since then integral equations have been largely hidden away and have been dominated by differential equations and the "big sister" FEM. It is an appropriate time to bring them out in the open.

4   NUMERICAL SOLUTION OF INTEGRAL EQUATIONS

The main task is now of course the solution of the integral equations. For practical problems only a numerical solution is possible. In a well posed boundary value problem either u is known (Dirchlet BC) or t is known (Neumann BC) on parts of the boundary and a solution is sought for the unknown.

In the following we discuss various methods for the solution of integral equations in a historical way starting with a method proposed by E.J. Nyström [6] in 1930.

4.1   Nyström method

The basic idea of the method is quite simple: We evaluate the integrals using Gauss Quadrature. This is explained on a flow example and a Neumann problem with known values of $t = t_0$ and the integral equation can be written as:

$$\int_S T(\mathbf{y}, \mathbf{x}) \cdot (u(\mathbf{x}) - u(\mathbf{y}))dS + 1 = \int_S U(\mathbf{y}, \mathbf{x}) \cdot t_0(\mathbf{x})dS \tag{61}$$

---

[5]Bemerkungen ueber $\Delta z = \frac{\partial^2 z}{\partial x^2} + \frac{\partial^2 z}{\partial y^2} = 0$, Journal fuer reine und angewandte Mathematik, Bd. 103.

To apply the integration rule we have to first introduce a local coordinate ξ that ranges from -1 to 1. For the integral on the left we have:

$$\int_S T(\mathbf{y}, \mathbf{x}) \cdot (u(\mathbf{x}) - u(\mathbf{y})) dS = \int_{-1}^{1} T(\mathbf{y}, \mathbf{x}) \cdot (u(\mathbf{x}) - u(\mathbf{y})) \cdot J \cdot d\xi \tag{62}$$

where J is the Jacobian of the transformation from $\mathbf{x}$ to ξ. Next we replace the integrals by a sum:

$$\int_{-1}^{1} T(\mathbf{y}, \mathbf{x}) \cdot (u(\mathbf{x}) - u(\mathbf{y})) \cdot J \cdot d\xi = \sum_{g=1}^{G} T(\mathbf{y}, \mathbf{x}_g) \cdot (u(\mathbf{x}_g) - u(\mathbf{y})) \cdot J_g \cdot W_g \tag{63}$$

where G is the number of Gauss points and W_g are weights. To determine the unknowns $u(\mathbf{x}_g)$ we need G equations. We now proceed in a similar way to the Trefftz method, i.e. we consider G different locations of the sources.

However instead of the sources being in different locations we assume that they are placed at the same points as the locations of the unknowns, i.e. the Gauss points. We obtain the following system of equations

$$\mathbf{T} \cdot \mathbf{u} = \mathbf{F} \tag{64}$$

where

$$\mathbf{T} = \begin{pmatrix} 1 - a_1 & T(\mathbf{y}_1, \mathbf{x}_2) \cdot J_2 \cdot W_2 & \cdots \\ T(\mathbf{y}_2, \mathbf{x}_1) \cdot J_2 \cdot W_2 & 1 - a_2 & \cdots \\ \cdots & \cdots & \cdots \end{pmatrix} \tag{65}$$

and

$$\mathbf{u} = \begin{pmatrix} u(\mathbf{x}_1) \\ u(\mathbf{x}_2) \\ \vdots \end{pmatrix}, \quad \mathbf{F} = \begin{pmatrix} F_1 \\ F_2 \\ \vdots \end{pmatrix} \tag{66}$$

with

$$a_n = \sum_{g=1}^{G} T(\mathbf{y}_n, \mathbf{x}_g) \cdot J_g \cdot W_g|_{g \neq n} \tag{67}$$

and

$$F_n = \int_S U(\mathbf{y}_n, \mathbf{x}) t_0(\mathbf{x}) dS \tag{68}$$

This integral can not be evaluated directly because of the singularity of the Kernel U. One way of evaluating the integral is to use a local correction or regularization.

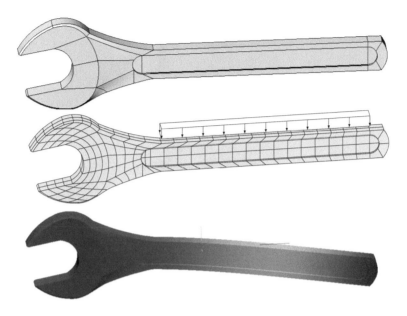

Figure 10 Example of the application of the Nyström method to the simulation of a spanner. Top: CAD model showing patches, Middle: Boundary conditions and subdivision into integration regions. Bottom: Displaced shape.

We replace the Kernel U by a *locally corrected* Kernel U^* that is defined by

$$U^*(\mathbf{y}_n, \mathbf{x}) = L(\mathbf{y}_n, \mathbf{x}) \quad \text{if } \mathbf{x} \in \Omega_y \tag{69}$$

$$U^*(\mathbf{y}_n, \mathbf{x}) = U(\mathbf{y}_n, \mathbf{x}) \quad \text{otherwise} \tag{70}$$

where Ω_y is a small region of exclusion around point $\mathbf{y}_n$. L can be computed by solving the following linear system of equations

$$\sum_{n=1}^{N} N_n(\mathbf{y}_n) L(\mathbf{y}_n, \mathbf{x}) = \int_{\Omega_y} U(\mathbf{y}_n, \mathbf{x}) N_n(\mathbf{y}_n) d\Omega \tag{71}$$

where $N_n(\mathbf{y}_n)$ are suitable test functions (polynomials).

A detailed explanation of the method is beyond the scope of this book but interested readers may consult an upcoming publication on this subject ([11]) where NURBS technology is applied for the first time. An example of application, namely the simulation of a spanner, is shown in Figure 10. Using NURBS patches for the description of the surface of the problem provides an added value because of the better control of continuity, an important aspect of this method.

Figure 11 Boris Galerkin.

4.2 Galerkin method

The Galerkin[6] method is actually an application of the residual method introduced earlier for the solution of differential equations. The aim is to minimize the error in the satisfaction of the integral equation over S.

We assume to have an approximation of $\mathbf{u} \approx \tilde{\mathbf{u}}$ and $\mathbf{t} \approx \tilde{\mathbf{t}}$ and want to obtain a solution that has the minimum error. We introduce *test functions* f_n into (55) and obtain N integral equations of the type:

$$\int_S f_n(\mathbf{y}) \int_S \mathbf{U}(\mathbf{y},\mathbf{x}) \cdot \tilde{\mathbf{t}}(\mathbf{x}) dS = \int_S f_n(\mathbf{y}) \int_S \mathbf{T}(\mathbf{y},\mathbf{x}) \cdot (\tilde{\mathbf{u}}(\mathbf{x}) - \tilde{\mathbf{u}}(\mathbf{y})) dS \tag{72}$$

$$n = 1, 2, \ldots, N$$

which can be solved for the unknowns. Test functions are usually chosen to be the basis functions for the approximation of the unknown. The method adds complexity to the implementation and requires additional numerical work, since a double (or triple) integral has to be solved numerically. It is popular with mathematicians since it lends itself easily to theoretical error analysis.

4.3 Collocation

In this method we ensure that the integral equation is only satisfied at a discrete number of points $\mathbf{y}_n$:

$$\int_S \mathbf{U}(\mathbf{y}_n,\mathbf{x}) \cdot \tilde{\mathbf{t}}(\mathbf{x}) dS = \int_S \mathbf{T}(\mathbf{y}_n,\mathbf{x}) \cdot (\tilde{\mathbf{u}}(\mathbf{x}) - \tilde{\mathbf{u}}(\mathbf{y}_n)) dS \quad n = 1, 2, \ldots, N \tag{73}$$

---

[6]Boris Grigoryevich Galerkin (1871 to 1945), born in Polotsk, Russian Empire, was a mathematician and an engineer.

This method avoids the additional integration, is easier to implement and requires less numerical work. On first glance it appears to be much less accurate, but experience showed that the loss in accuracy is small and easily compensated by the decrease in complexity of implementation and run times.

4.4 Discretisation

Whether we use Galerkin or Collocation we need to discretize the integral equations in order to solve them. As with the FEM it is convenient to divide the integration region into subregions and to assume a piecewise approximation of $\mathbf{u}$ and $\mathbf{t}$ over it. For the collocation method the integral equation can be re-written as

$$\sum_{e=1}^{E}\int_{S}\mathbf{U}(\mathbf{y}_n,\mathbf{x})\cdot\tilde{\mathbf{t}}(\mathbf{x})dS_e=\sum_{e=1}^{E}\int_{S}\mathbf{T}(\mathbf{y}_n,\mathbf{x})\cdot(\tilde{\mathbf{u}}(\mathbf{x})-\tilde{\mathbf{u}}(\mathbf{y}_n))dS_e \tag{74}$$

$$n=1,2,\ldots,N$$

where E is the number of subregions (elements). One of the early implementation of discretisation appeared in [7].

In the simplest approach we assume that the values of $\tilde{\mathbf{t}}$ and $\tilde{\mathbf{u}}$ are constant within an element e which means that they can be taken outside the integral. We assume that the source points $\mathbf{y}_n$ are located at the centers of the elements. The discretized integral equations can now be written as:

$$\sum_{e=1}^{E}\tilde{\mathbf{t}}^e\cdot\int_{S}\mathbf{U}(\mathbf{y}_n,\mathbf{x})dS_e=\sum_{e=1}^{E}(\tilde{\mathbf{u}}^e-\tilde{\mathbf{u}}^n)\int_{S}\mathbf{T}(\mathbf{y}_n,\mathbf{x})dS_e \quad n=1,2,\ldots,N \tag{75}$$

where $\tilde{\mathbf{t}}^e$ and $\tilde{\mathbf{u}}^e$ are the approximate values of $\mathbf{t}$ and $\mathbf{u}$ inside element e.

Let us demonstrate the method on the example of the circular insulator, solved with the Trefftz method. Figure 12 shows a possible discretization of the insulator into constant elements.

The discretized integral equation for this problem is

$$\sum_{e=1}^{E}\Delta U_n^e\cdot t^e=\sum_{e=1}^{E}\Delta T_n^e(u^e-u^n) \tag{76}$$

where

$$\Delta U_n^e=\int_{S_e}U(\mathbf{y}_n,\mathbf{x})dS_e; \quad \Delta T_n^e=\int_{S_e}T(\mathbf{y}_n,\mathbf{x})dS_e \tag{77}$$

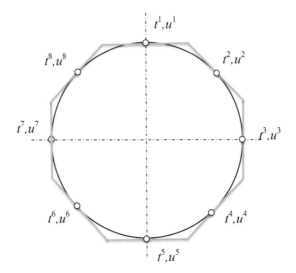

Figure 12 Flow past insulator: Discretisation into constant elements.

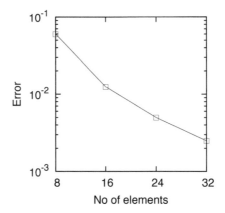

Figure 13 Flow past insulator: Convergence of maximum potential.

Because the integration regions are straight lines the integrals can be evaluated analytically (for details see [2]). Equation (76) can be written as a system of equations

$$\mathbf{Tu} = \mathbf{Ut}_0 \tag{78}$$

and with the values of $\mathbf{t}_0$ known, can be solved for the unknown potentials $\mathbf{u}$.

Figure 13 shows the convergence of the solution. It can be seen that the convergence is much slower than for the Trefftz method. This is hardly surprising since the approximation of the geometry and the known and unknown values is quite crude.

5   SUMMARY AND CONCLUSIONS

We have introduced the concept of using fundamental solutions of the differential equation. The benefits are immediately apparent: Since the solution inside the domain now satisfies the differential equation exactly, no approximation occurs there and the only approximation is in the satisfaction of the boundary conditions.

Historically, the first attempt at a solution with this idea was performed by Trefftz, who proposed it as an alternative to the Ritz method, introduced earlier. Unfortunately this method is not usable for practical applications, mainly due to the fact that source points need to be specified.

As an alternative we introduced integral equations. Using well known engineering principles, such Betti's theorem, we derived the Somigliana identity in a much easier and understandable way than proposed by mathematicians (see for example the Appendix in [3] where the derivation is made by applying Green's theorem). Using a limiting procedure we managed to move the source points to the boundary thereby making the method usable. The result is an integral equation linking the known and unknown value on the boundary.

The integral equations can be solved by discretization, i.e. by dividing the boundary into smaller elements and assuming an approximation over each element much in the same way as is done in the finite element method. Two possibilities have been proposed for the solution: One involving a minimization of the error in satisfying the integral equation at all points on the boundary (Galerkin method), the other involving a satisfaction only at discrete points (Collocation method). Mathematicians favor the former since error estimates can be derived. However, engineers have found that the additional effort in implementation and in the solution does not lead to a significant increase in accuracy. Therefore Collocation has been the favorite by engineers for practical implementation. Indeed, the first papers on the implementation of isogeometric BEM technology ([8], [9], [5] and [1]) use Collocation. We will follow this trend here.

The first benefit of boundary integral approach becomes apparent immediately: The dimension of the problem has been reduced. This means that we only have to deal with values at the boundary of the domain, with the values inside the domain computed using functions that satisfy the differential equation exactly.

In elasticity this means for example that equilibrium and compatibility are satisfied exactly inside the domain. This results in a second benefit, namely that we expect more accurate results inside the domain, than with domain methods such as the Finite Element Method.

Finally, this method is ideally suited to the isogeometric concept and means that data from CAD programs, that describe surfaces, can be used directly for the simulation. This comes at a price, however. Solving the integral equations is not an easy task, considering that the fundamental solutions are singular. This will be a major challenge in the implementation, discussed next. Furthermore, all methods lead to fully populated system matrices, another difference to the FEM. For the method to be competitive therefore the number of unknowns has to be significantly smaller.

Comparison of the number of unknowns with domain methods depends on the ratio of boundary surface to volume. For problems such as tunnels and underground caverns, the volume is much greater than the boundary surface (i.e. surface of the excavation) and for practical purposes the volume of the ground can be assumed to be infinite. Therefore, for these types of problems, the method will always be superior to domain methods. This is why the first practical uses of the method occurred in mining ([4]). For other problems, involving low volume to surface ratios, the method will only be competitive if the quality of the results is better. The last example showed that using a constant approximation over linear segments provides poor quality results. So obviously the aim has to use higher order approximations. Indeed, the use of NURBS is ideal for this purpose.

Therefore we introduce the Boundary Element Method next.

BIBLIOGRAPHY

[1]   G. Beer. Mapped infinite patches for the NURBS based boundary element analysis in geomechanics. *Computers and Geotechnics*, 66:66–74, 2015.

[2]   G. Beer, I. Smith and C. Duenser. *The Boundary Element Method with Programming*. Springer-Verlag, Wien, 2008.

[3]   G. Beer and J.O. Watson. *Introduction to Finite and Boundary Element Methods for Engineers*. Wiley, 1992.

[4]   F.H. Deist, M.D.G. Salamon and E. Georgiadis. A new digital method for three-dimensional stress analysis in elastic media. *Rock Mechanics*, 5(189–202), 1973.

[5]   Benjamin Marussig, Jürgen Zechner, Gernot Beer and Thomas-Peter Fries. Fast isogeometric boundary element method based on independent field approximation. *Computer Methods in Applied Mechanics and Engineering*, 284(0):458–488, 2015. Isogeometric Analysis Special Issue.

[6]   E.J. Nyström. Über die praktische Auflösung von Integral-gleichungen mit Anwendungen auf Randwertaufgaben. *Acta Math.*, 54:185–204, 1930.

[7]   F.J. Rizzo. An integral equation approach to boundary value problems in classical elastostatics. *Q. Appl. Math.*, 25:83–95, 1967.

[8]   M.A. Scott, R.N. Simpson, J.A. Evans, S. Lipton, S.P.A. Bordas, T.J.R. Hughes and T.W. Sederberg. Isogeometric boundary element analysis using unstructured T-splines. *Computer Methods in Applied Mechanics and Engineering*, 254(0):197–221, 2013.

[9]   R.N. Simpson, S.P.A. Bordas, J. Trevelyan and T. Rabczuk. A two-dimensional isogeometric boundary element method for elastostatic analysis. *Computer Methods in Applied Mechanics and Engineering*, 209–212(0):87–100, February 2012.

[10]   Erich Trefftz. Ein Gegenstück zum Ritzschen Verfahren. In *Proc. 2. Int. Congress in Applied Mechanics*, 1926.

[11]   J. Zechner, B. Marussig, G. Beer and T.P. Fries. The Isogeometric Nyström Method. *Computer Methods in Applied Mechanics and Engineering*, submitted 2015.

Chapter 8

Stage 7: The boundary element method for plane problems

There is nothing more powerful than an idea whose time has come

V. Hugo

where we unveil the beauty of Cinderella, so that it can be seen.

1 INTRODUCTION

In the previous stage we discovered that using integral equations brings us closer to CAD, because the problem is defined by boundary curves or surfaces. We learned that the integral equations can be solved by an approximation of the boundary values, but that assuming constant variation does not yield good results. Lagrange/Serendipity functions, as are being used in the FEM seem an obvious choice for approximating the boundary values. This was first suggested by Lachat and Watson [2] for the description of the geometry as well as the approximation of the boundary values. In the explanation of the boundary element method (BEM) we start with plane problems and the classical isoparametric approach, using Serendipity functions. We then introduce NURBS and highlight the salient differences. Three-dimensional problems are discussed next. We will see that the most challenging part of the implementation is the evaluation of the integrals due to the singular nature of the Kernels.

Notation A few more words are appropriate here on the notation used. We define by u or $\mathbf{u}$ the potential or displacement vector and by t or $\mathbf{t}$ the flow or traction vector. Note the difference to the local coordinates u and t (not italic). The parameters are defined locally for NURBS patch e by u_i^e or $\mathbf{u}_i^e$ or t_i^e or $\mathbf{t}_i^e$. The global parameters are specified as u_i or $\mathbf{u}_i$ or t_i or $\mathbf{t}_i$. In the following we use the notation for vectors, that would revert to scalars for potential problems. We refer to the NURBS curves, that describe geometry, as patches.

2 CLASSICAL ISOPARAMETRIC APPROACH

For each boundary element e the geometry and boundary values are approximated by:

$$\mathbf{x}^e(\xi) = \sum_{i=1}^{I} N_i(\xi) \cdot \mathbf{x}_i^e \tag{1}$$

$$\mathbf{u}^e(\xi) = \sum_{i=1}^{I} N_i(\xi) \cdot \mathbf{u}_i^e \tag{2}$$

$$\mathbf{t}^e(\xi) = \sum_{i=1}^{I} N_i(\xi) \cdot \mathbf{t}_i^e \tag{3}$$

where N_i are suitable basis functions of the local coordinate ξ, $\mathbf{x}_i^e$ specify the location of the nodal points, $\mathbf{u}_i^e$, $\mathbf{t}_i^e$ are nodal values of $\mathbf{u}$, $\mathbf{t}$ and I is the number of element nodes. The term isoparametric means that the same functions are used for all descriptions.

The collocation points $\mathbf{y}_n$ are conveniently taken to be the nodal points of elements. The discretized integral equation can be written as:

$$\sum_{e=1}^{E} \int_{S_e} \mathbf{U}(\mathbf{y}_n, \mathbf{x}) \cdot \mathbf{t}(\mathbf{x}) dS_e = \sum_{e=1}^{E} \left[\int_{S_e} \mathbf{T}(\mathbf{y}_n, \mathbf{x}) \cdot \mathbf{u}(\mathbf{x}) dS_e - \mathbf{u}(\mathbf{y}_n) \int_{S_e} \mathbf{T}(\mathbf{y}_n, \mathbf{x}) dS_e \right]$$
$$n = 1, 2, \ldots, N \tag{4}$$

Note that because $\mathbf{u}(\mathbf{y}_n)$ is constant, it can be taken outside the integral.

Substitution of the approximations and changing the integration limits to $-1, +1$ we obtain:

$$\sum_{e=1}^{E} \int_{-1}^{1} \mathbf{U}(\mathbf{y}_n, \mathbf{x}^e(\xi)) \left[\sum_{i=1}^{I} N_i(\xi) \, \mathbf{t}_i^e \right] J \, d\xi$$

$$= \sum_{e=1}^{E} \int_{-1}^{1} \mathbf{T}(\mathbf{y}_n, \mathbf{x}^e(\xi)) \left[\sum_{i=1}^{I} N_i(\xi) \mathbf{u}_i^e \right] J \, d\xi - \left[\sum_{i=1}^{I} N_i(\xi_n) \mathbf{u}_i^e \right] \sum_{e=1}^{E} \int_{-1}^{1} \mathbf{T}(\mathbf{y}_n, \mathbf{x}^e(\xi)) J \, d\xi$$
$$n = 1, \ldots, N \tag{5}$$

where N is the total number of collocation points ($=$ nodal points) and ξ_n is the local coordinate of collocation point n.

The nodal values can be taken outside the integrals resulting in:

$$\sum_{e=1}^{E} \sum_{i=1}^{I} \left[\int_{-1}^{1} \mathbf{U}(\mathbf{y}_n, \mathbf{x}^e(\xi)) N_i(\xi) J \, d\xi \right] \mathbf{t}_i^e = \sum_{e=1}^{E} \sum_{i=1}^{I} \left[\int_{-1}^{1} \mathbf{T}(\mathbf{y}_n, \mathbf{x}^e(\xi)) \cdot [N_i(\xi)] J \, d\xi \right] \mathbf{u}_i^e$$

$$- \left[\sum_{i=1}^{I} N_i(\xi_n) \mathbf{u}_i^{en} \right] \sum_{e=1}^{E} \int_{-1}^{1} \mathbf{T}(\mathbf{y}_n, \mathbf{x}^e(\xi)) J \, d\xi$$
$$\tag{6}$$

where *en* indicates an element that contains the collocation point n.

Changing left and right hand side, a set of equations for N nodal points can be obtained:

$$\sum_{e=1}^{E}\sum_{i=1}^{I}\Delta\mathbf{T}_{ni}^{e}\mathbf{u}_{i}^{e} - \left[\sum_{i=1}^{I}N_{i}(\xi_{n})\mathbf{u}_{i}^{en}\right]\mathbf{T}_{n} = \sum_{e=1}^{E}\sum_{i=1}^{I}\Delta\mathbf{U}_{ni}^{e}\mathbf{t}_{i}^{e} \quad \text{for } n=1,2,3,\ldots,N \quad (7)$$

with

$$\Delta\mathbf{U}_{ni}^{e} = \int_{-1}^{1}\mathbf{U}\left(\mathbf{y}_{n},\mathbf{x}^{e}(\xi)\right)N_{i}(\xi)J\,d\xi \tag{8}$$

$$\Delta\mathbf{T}_{ni}^{e} = \int_{-1}^{1}\mathbf{T}\left(\mathbf{y}_{n},\mathbf{x}^{e}(\xi)\right)\cdot N_{i}(\xi)J\,d\xi \tag{9}$$

$$\mathbf{T}_{n} = \sum_{e=1}^{E}\int_{-1}^{1}\mathbf{T}\left(\mathbf{y}_{n},\mathbf{x}^{e}(\xi)\right)J\,d\xi \tag{10}$$

Equation (7) can be assembled into a system of equations

$$[\mathbf{T}]\{\mathbf{u}\} = [\mathbf{U}]\{\mathbf{t}\} \tag{11}$$

where $\{\mathbf{u}\}$ and $\{\mathbf{t}\}$ contain the values of $\mathbf{u}$ and $\mathbf{t}$ at all nodes. Matrices are assembled in a loop over all elements and nodes

$$\mathbf{T}_{n,inci(i)} = \mathbf{T}_{n,inci(i)} + \Delta\mathbf{T}_{ni}^{e} \tag{12}$$

$$\mathbf{U}_{n,inci(i)} = \mathbf{U}_{n,inci(i)} + \Delta\mathbf{U}_{ni}^{e}$$

where $inci(i)$ is the global node number of the local node i (see Figure 1). The Kroneker Delta property of the basis function means that diagonal sub-matrices need not be computed. Instead they are determined by taking the negative sum of the off-diagonal sub-matrices. In the case of an infinite or semi-infinite domain problem the value of the azimuthal integral has to be added as explained previously.

2.1   Numerical evaluation of integrals

Here we discuss the numerical evaluation of the integrals. We divide the integrals into the following cases:

- **Regular integrals:** This is the case when the collocation point is not inside element e or − in the case it is − the integrand involves the Kernel U and the shape function tends to zero as collocation point is approached, therefore canceling out the singularity.
- **Nearly singular integrals:** This is the same as for regular integrals, except that the collocation point is near the element being integrated.
- **Weakly singular integrals:** This is the case where the collocation point is inside the element, the shape function does not tend to zero as the point is approached and the integrand involves the Kernel U.

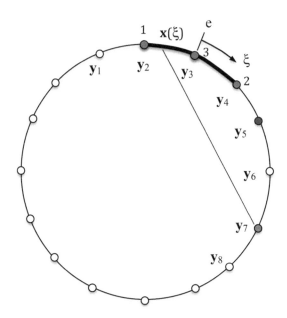

Figure I A plane problem showing collocation points and discretization with quadratic Serendipity boundary elements. One boundary element is highlighted together with its nodal points, local coordinate and local numbering (*inci* in this case is (2, 4, 3)). Collocation points marked in red indicate singular integration, in blue nearly singular integration and in green regular iteration.

- **Strongly singular integrals:** This is the case where the collocation point is inside the element and the integrand involves the Kernel $\mathbf{T}$. This integral only exists as Cauchy principal value. However, by introducing the rigid body trick and considering the Kroneker Delta property of the basis function, this integral need not be computed.

Regular and nearly singular integration   For this case we can employ Gauss Quadrature. The regular integrals are given by

$$\Delta U_{ni}^e = \sum_{m=1}^{M} \mathbf{U}\big(\mathbf{y}_n, \mathbf{x}^e(\xi_m)\big)\, N_i(\xi_m) J(\xi_m) \cdot W_m$$

$$\Delta T_{ni}^e = \sum_{m=1}^{M} \mathbf{T}\big(\mathbf{y}_n, \mathbf{x}^e(\xi_m)\big)\, N_i(\xi_m) J(\xi_m) \cdot W_m \tag{13}$$

$$\mathbf{T}_n = \sum_{e=1}^{E} \sum_{m=1}^{M} \mathbf{T}\big(\mathbf{y}_n, \mathbf{x}^e(\xi_m)\big)\, J(\xi_m) \cdot W_m$$

In the above M is the number of Gauss points (chosen depending on the proximity of the collocation point to the element), ξ_m are coordinates of Gauss points, J is the

Table I   Required number of Gauss points.

R/L	≮1.4025	≮0.6736
M	3	4

Jacobian of the transformation between the local and the global coordinate system and W_m are weights. For nearly singular integrals the collocation point is near the element and the Kernel function rises sharply. If the number of integration points is not increased considerably the results will be in error.

Early work of adapting the Gauss Quadrature method to the isoparametric BEM was carried out by Watson [6]. By only considering the singular part of the Kernel he managed to develop approximate formulae for the integration error, based on error estimates by Stroud and Secrest [5]. For example, integrating the function $\frac{1}{r}$ in the local coordinate ξ space, ranging from -1 to 1, the approximate integration error as a function of the number of Gauss points M can be written as:

$$\epsilon = \left| \int_{-1}^{1} \frac{1}{r(\xi)} d\xi - \sum_{m=1}^{M} W_m \frac{1}{r(\xi_m)} \right| \leqslant \frac{L}{4}^{2M} \frac{4}{R^{2M+1}} \tag{14}$$

where L is the length of the integration region and R is the shortest distance of the collocation point to the integration region. This error indicator can be used to determine the number of integration points depending on L and R (see [1]). Regarding the maximum number of Gauss points that should be used, it has been found that it is more efficient to use few points and to subdivide the integration region.

Table 1 shows the number of Gauss points required for the integration involving a Kernel of $o(\frac{1}{r})$ depending on the value of R/L not being smaller than a certain value. We can use this criterion also for the Kernel with $o(\ln\frac{1}{r})$ and be on the safe side. For values of $\frac{R}{L} < 0.6736$ the integration region has to be subdivided.

Weakly singular integrals   The kernel U has a singularity of $o(\ln(\frac{1}{r}))$. The product UN_i tends to infinity for cases where N_i does not tend to zero at the singularity point.

For this we can use a coordinate transformation between the coordinate ξ and a coordinate γ as originally proposed by Telles [3] resulting in a zero Jacobian at the collocation point, thereby canceling the singularity. This means that the Integration in the γ coordinate space can be carried out using normal Gauss integration.

For example, if the collocation point is at $\gamma = -1$, the transformation is given by:

$$\xi(\gamma) = \frac{1}{2}(\gamma^2 - 1) + \gamma + 1 \tag{15}$$

The Jacobian of this transformation is $J(\gamma) = 1 + \gamma$ and approaches 0 for $\gamma = -1$.

The Integration can now be carried out using standard Gauss Quadrature

$$\Delta U_{ni}^e = \int_{-1}^{1} U(\mathbf{y}_n, \mathbf{x}^e(\xi(\gamma))) N_i(\xi(\gamma)) \cdot J \cdot J(\gamma) \cdot d\gamma \tag{16}$$

$$= \sum_{m=1}^{M} U(\mathbf{y}_n, \mathbf{x}^e(\xi(\gamma_m))) N_i(\xi(\gamma_m)) \cdot J \cdot J(\gamma_m) \cdot W_m$$

3 NURBS BASED APPROACH

If we use a NURBS based approach subtle changes have to be made in the implementation. The first action will be to depart from the isoparametric concept by allowing different approximations for the geometry and the variation of the boundary values:

$$\mathbf{x}^e = \sum_{i=1}^{I} N_i(\mathbf{u}) \cdot \mathbf{x}_i^e \tag{17}$$

$$\mathbf{u}^e = \sum_{i=1}^{I^d} N_i^d(\mathbf{u}) \cdot \mathbf{u}_i^e \tag{18}$$

$$\mathbf{t}^e = \sum_{i=1}^{I^t} N_i^t(\mathbf{u}) \cdot \mathbf{t}_i^e \tag{19}$$

where $N_i(\mathbf{u})$ are the basis functions for approximating the geometry and $\mathbf{x}_i^e$ are the coordinates of the control points. $N_i^d(\mathbf{u})$ are the basis functions for describing the variation of $\mathbf{u}$ and $\mathbf{u}_i^e$ are the corresponding parameter values. Finally $N_i^t(\mathbf{u})$ are the basis functions for describing the variation of $\mathbf{t}$ and $\mathbf{t}_i^e$ are the corresponding parameter values. Recall that $\mathbf{u}$ and $\mathbf{t}$ are either two-dimensional vectors in the case of elasticity or scalars in the case of potential problems.

The next thing that is different to the previous approach is that we no longer have nodal points that can be used as collocation points. Therefore the location of these points has to be explicitly computed. Since we need as many collocation points as unknown parameters it is logical to put these points where the anchors of the basis functions are. To compute the location of collocation points we determine their location first in local u coordinates using the Greville formula introduced earlier and then in the global x, y coordinates. This is explained in Figure 2 showing a possible discretization of a circular boundary. Two NURBS patches are shown with control points and the outward normal. This geometry is defined with the basis functions $N_i(\mathbf{u})$ shown under b). The variation of $\mathbf{u}$ is defined by the basis functions $N_i^d(\mathbf{u})$ shown after refinement under c) together with the anchors. These anchors are used to compute the collocation points. The local coordinates of the anchors are:

$$u_{i(n)} = \frac{u_{i+1} + u_{i+2} + \ldots + u_{i+p}}{p} \quad i = 0, 1, \ldots, I \tag{20}$$

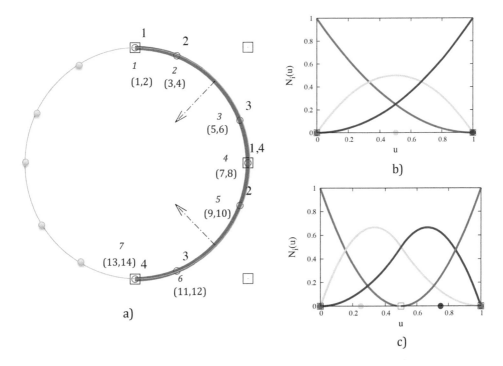

Figure 2 A circular boundary with a) 2 NURBS patches with outward normal, control points (squares) and collocation points (circles) showing local and global numbering of parameters for potential and elasticity problems (in parentheses) b) basis functions used for the description of the geometry c) basis functions for the description of the unknown **u** (the anchors shown by filled circles are used to compute the location of collocation points).

where $i(n)$ is the local number of the collocation point n and $u_{i+1} \dots$ are the entries in the Knot vector.

The global coordinates are computed by:

$$\mathbf{y}_n = \sum_{i=1}^{I} N_i(\mathbf{u}_{i(n)}) \cdot \mathbf{x}_i^e \tag{21}$$

The discretized BEM equations are now:

$$\sum_{e=1}^{E}\sum_{i=1}^{I} \Delta T_{ni}^e \mathbf{u}_i^e - \left[\sum_{i=1}^{I} N_i(\mathbf{u}_n)\mathbf{u}_i^{en} \right] \mathbf{T}_n = \sum_{e=1}^{E}\sum_{i=1}^{I} \Delta U_{ni}^e \mathbf{t}_i^e \quad \text{for } n = 1, 2, 3, \dots, N \tag{22}$$

An important point is that the basis functions are not necessarily zero at the collocation points, as has been the case for the isoprametric BEM. This can be seen for the middle collocation points in Figure 2c, where all shape functions have nonzero values. Therefore the application of the rigid body trick is more complicated.

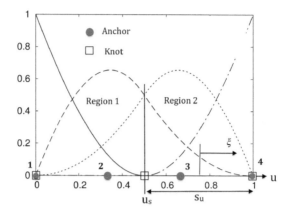

Figure 3 Plot along the NURBS patch of Figure 2 in the *u* coordinate system with basis functions for the description of the unknown showing the subdivision of the integration region.

Regular integration We recall that in the case where the basis functions for the description of **u** or **t** have multiple Knot entries the NURBS patch has to be subdivided into integration regions because some functions have limited span.

In addition, to apply Gauss Quadrature we have to change from the NURBS coordinate u to coordinate ξ, ranging from -1 to $+1$. The mapping between the ξ and u coordinates for subregion s is given by:

$$u = \frac{s_s}{2}(\xi + 1) + u_s \tag{23}$$

where s_s is the size of the subregion and u_s is the local coordinate of the start of the subregion (see Figure 3).

The Jacobian of this transformation is

$$\frac{du}{d\xi} = \frac{s_s}{2} \tag{24}$$

The regular integrals are given by

$$\Delta U_{ni}^e = \sum_{s=1}^{S}\sum_{m=1}^{M} U(\mathbf{y}_n, \mathbf{x}^e(u_m))\, N_i^t(u_m) J(u_m)\frac{\partial u}{\partial \xi} \cdot W_m \tag{25}$$

$$\Delta T_{ni}^e = \sum_{s=1}^{S}\sum_{m=1}^{M} T(\mathbf{y}_n, \mathbf{x}^e(u_m))\, N_i^d(u_m) J(u_m)\frac{\partial u}{\partial \xi} \cdot W_m$$

$$T_n^e = \sum_{e=1}^{E}\sum_{m=1}^{M} T(\mathbf{y}_n, \mathbf{x}^e(u_m))\,\frac{1}{2} \cdot W_m$$

where S is the number of subregions. Note that the last integral does not require a subdivision, since no basis function is involved.

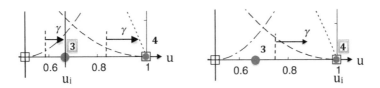

Figure 4 Explanation of the coordinate transformation between γ and u, left for an arbitrary position of the collocation point, right where the collocation point is at an edge.

Weakly singular integrals   As before we use a coordinate transformation between the NURBS coordinate u and a coordinate γ with the limits -1 and $+1$ resulting in a zero Jacobian at the collocation point, thereby canceling the singularity.

If the collocation point is at an arbitrary position u_i inside the integration region, it has to be subdivided into 2 sub-regions as shown in Figure 4. The transformation for sub-region 1 is given by

$$u(\gamma) = \frac{u_i - u_1}{2}\left(\frac{1}{2}(1 - \gamma^2) + \gamma + 1\right) + u_1 \tag{26}$$

The Jacobian of this transformation is $J(\gamma) = (u_i - u_1)/2 \cdot (1 - \gamma)$ and tends to zero for $\gamma = 1$. For subregion 2 the transformation is

$$u = \frac{u_2 - u_i}{2}\left(\frac{1}{2}(\gamma^2 - 1) + \gamma + 1\right) + u_i \tag{27}$$

The Jacobain of this transformation is $J(\gamma) = (u_2 - u_i)/2 \cdot (1 + \gamma)$ and tends to zero for $\gamma = -1$.

For the case where the collocation points are on the edges of the element the coordinate transformation is given by for the collocation point at $\gamma = -1$

$$u(\gamma) = \frac{s_u}{2}\left(\frac{1}{2}(\gamma^2 - 1) + \gamma + 1\right) + u_1 \tag{28}$$

The Jacobian of this transformation is $J(\gamma) = s_u/2 \cdot (1 + \gamma)$ and approaches 0 for $\gamma = -1$. For the collocation point at $\gamma = 1$ we have

$$u(\gamma) = \frac{s_u}{2}\left(\frac{1}{2}(1 - \gamma^2) + \gamma + 1\right) + u_1 \tag{29}$$

The Jacobian of this transformation is $J(\gamma) = s_u/2 \cdot (1 - \gamma)$ and approaches 0 for $\gamma = 1$.
The Integration can now be carried out using standard Gauss Quadrature

$$\Delta U^e_{n,i} = \sum_{sr=1}^{S} \int_{-1}^{1} U(y_n, x(u(\gamma)))N_i(u(\gamma))\, J(u)J(\gamma)d\gamma \tag{30}$$

$$= \sum_{sr=1}^{S}\sum_{g=1}^{M} U(y_n, x(u(\gamma_m)))N_i^t(u(\gamma_m))\, J(u(\gamma_m))J(\gamma_m)\, W_m \tag{31}$$

where S is the number of subregions and M is the number of Gauss points.

3.1 Boundary conditions

Since the primary unknowns are not nodal point values but parameters, which have no physical meaning, the imposition of Boundary conditions needs some consideration and this will be discussed here.

Dirichlet BC   The value of **u** at a particular location u_d in a NURBS patch is given by

$$\mathbf{u}(u_d) = \sum_{i=1}^{I_d} N_i^d(u_d) \cdot \mathbf{u}_i \tag{32}$$

If $\mathbf{u}(u_d)$ is specified as zero then all parameters $\mathbf{u}_i$ for basis functions that are not zero at u_d must be set to zero. For a non zero value the parameters have to be determined. For example for a specified value $\mathbf{u}_d$ that is constant along the patch we assign this value to all anchors and have:

$$\mathbf{u}(u) = \mathbf{u}_d \sum_{i=1}^{I_d} N_i^d(u) \tag{33}$$

and with the partition of unity condition $\sum_{i=1}^{I_d} N_i^d(u) = 1$ the computed parameter value is $\mathbf{u}_d$ along the patch.

Neumann BC   If the boundary condition is a distributed load **q**, it contributes to the right hand side of the equations as:

$$\mathbf{F}_n = \sum_{s=1}^{S} \sum_{m=1}^{M} \mathbf{U}(\mathbf{y}_n, \mathbf{x}^e(u_m)) \mathbf{q}(u_m) J(u_m) \frac{\partial u}{\partial \xi} \cdot W_m \tag{34}$$

Robin BC   In the case of the problem of the flow past an isolator, discussed previously, the Neumann Boundary value depends on the outward normal **n**. For this problem the right hand side contribution is computed by:

$$\mathbf{F}_n = \sum_{s=1}^{S} \sum_{m=1}^{M} \mathbf{U}(\mathbf{y}_n, \mathbf{x}^e(u_m)) \mathbf{q}_0 \cdot \mathbf{n}(u_m) J(u_m) \frac{\partial u}{\partial \xi} \cdot W_m \tag{35}$$

where $\mathbf{q}_0 = $ is the specified flow vector.

For the problem of an excavation in elasticity, the value depends on the outward normal and the virgin stress:

$$\mathbf{F}_n = \sum_{s=1}^{S} \sum_{m=1}^{M} \mathbf{U}(\mathbf{y}_n, \mathbf{x}^e(u_m)) \mathbf{t}_0 \cdot J(u_m) \frac{\partial u}{\partial \xi} \cdot W_m \tag{36}$$

where

$$\mathbf{t}_0 = \begin{pmatrix} \sigma_{x0} \cdot n_x + \tau_{xy0} \cdot n_y \\ \sigma_{y0} \cdot n_y + \tau_{xy0} \cdot n_x \end{pmatrix} \tag{37}$$

and $\sigma_{x0}, \sigma_{y0}, \tau_{xy0}$ are virgin stress components.

4   ASSEMBLY OF MULTIPLE PATCHES

So far the discussion was restricted to problems that can be described by one NURBS patch. Here we extend the discussion to problems involving multiple patches. At the location where the patches connect we usually expect a continuity of $\mathbf{u}$ (displacement or potential).

With regard to the continuity of $\mathbf{t}$ (traction or flow) this would depend on the problem. If it is a pure Neumann problem i.e. all values of $\mathbf{t}$ are known and there is no requirement for continuity.

4.1   Pure Neumann problem

A pure Neumann problem is for example an excavation in an infinite domain where $\mathbf{t}$ can be computed from the virgin stress field. The system of equations can be written as:

$$\sum_{e=1}^{E}\sum_{i=1}^{I}\Delta T_{ni}^{e}\mathbf{u}_{i}^{e} - \left[\sum_{i=1}^{I}N_{i}^{d}(\mathbf{u}_{n})\mathbf{u}_{i}^{en}\right]\mathbf{T}_{n} = \sum_{e=1}^{E}\Delta F_{n}^{e} \quad n=1,2,\ldots,N \tag{38}$$

which we have to assemble into a system of equations

$$[\mathbf{T}]\{\mathbf{u}\} = \{\mathbf{F}\} \tag{39}$$

Regarding the subtraction of the second term in the left hand side we note that for collocation point 1 only $N_{1}^{d}(\mathbf{u}_{1})$ is non-zero, whereas for collocation point 2 only $N_{1}^{d}(\mathbf{u}_{2})$, $N_{2}^{d}(\mathbf{u}_{2})$ and $N_{3}^{d}(\mathbf{u}_{2})$ are non-zero. The assembly of the left hand side can be sketched as follows:

$$[\mathbf{T}] = \begin{pmatrix} & 1 & 2 & 3 & 4 & \cdots \\ 1 & \Delta T_{11}^{1} & \Delta T_{12}^{1} & \Delta T_{13}^{1} & \Delta T_{14}^{1} + \Delta T_{11}^{2} & \cdots \\ & -N_{1}^{d}(\mathbf{u}_{1})\cdot\mathbf{T}_{1} & & & & \\ 2 & \Delta T_{21}^{1} & \Delta T_{22}^{1} & \Delta T_{23}^{1} & \Delta T_{24}^{1} + \Delta T_{21}^{2} & \cdots \\ & -N_{1}^{d}(\mathbf{u}_{2})\cdot\mathbf{T}_{2} & -N_{2}^{d}(\mathbf{u}_{2})\cdot\mathbf{T}_{2} & -N_{3}^{d}(\mathbf{u}_{2})\cdot\mathbf{T}_{2} & & \\ \vdots & \vdots & \vdots & \vdots & \vdots & \vdots \end{pmatrix} \tag{40}$$

4.2   Mixed Neumann/Dirichlet problem

In some cases we have a mixture of boundary conditions. Consider the cantilever beam in Figure 5 which has a mixture of Dirichlet and Neumann BCs.

In this case the system of equations can be written as

$$[\mathbf{A}]\{\mathbf{d}\} = \{\mathbf{F}\} \tag{41}$$

where $\{\mathbf{d}\}$ contains a mixture of $\mathbf{u}$ and $\mathbf{t}$ terms.

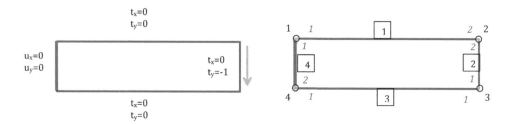

Figure 5 Left: Cantilever beam with Dirichlet BC along red boundary and Neumann BC along blue boundary. Right: Discretisation into 4 linear NURBS patches (numbers in boxes) with collocation points, global (black numbers) and local numbering of anchors (blue italic numbers).

For the cantilever beam in Figure 5 the assembly of matrix $\mathbf{A}$ results in:

$$
\mathbf{A} = \begin{pmatrix}
1 & 2 & 3 & 4 \\
-\Delta\mathbf{U}_{12}^4 & \Delta\mathbf{T}_{12}^1 + \Delta\mathbf{T}_{12}^2 & \Delta\mathbf{T}_{11}^2 + \Delta\mathbf{T}_{11}^3 & -\Delta\mathbf{U}_{12}^4 \\
-\Delta\mathbf{U}_{22}^4 & \begin{matrix} \Delta\mathbf{T}_{22}^1 + \Delta\mathbf{T}_{22}^2 \\ -N_2^{\bar{d}}(\mathbf{u}_2) \cdot \mathbf{T}_2 \end{matrix} & \Delta\mathbf{T}_{21}^2 + \Delta\mathbf{T}_{21}^3 & -\Delta\mathbf{U}_{22}^4 \\
\vdots & \vdots & \vdots & \vdots
\end{pmatrix}
\tag{42}
$$

The vector $\mathbf{F}$ is given by

$$
\mathbf{F} = \begin{pmatrix}
(\Delta\mathbf{U}_{11}^2 + \Delta\mathbf{U}_{12}^2)\mathbf{t}_0 \\
(\Delta\mathbf{U}_{21}^2 + \Delta\mathbf{U}_{22}^2)\mathbf{t}_0 \\
\vdots
\end{pmatrix}
\tag{43}
$$

where

$$
\mathbf{t}_0 = \begin{pmatrix} 0 \\ -1 \end{pmatrix}
\tag{44}
$$

4.3 Symmetry

In contrast to the imposition of symmetry in the FEM, symmetry conditions cannot be imposed by boundary conditions. However, we can take advantage of symmetry by only inputing the symmetric part of the boundary discretization, by constructing the whole boundary discretization internally in the program and using the fact that if unknowns are known on one side of the symmetry plane they are also known on the other side.

Consider the problem in Figure 2 but now with the y-axis and x-axis being axes of symmetry. We define the problem with only one patch (numbered 1) and mirror it three times across the symmetry planes (1',1",1"').

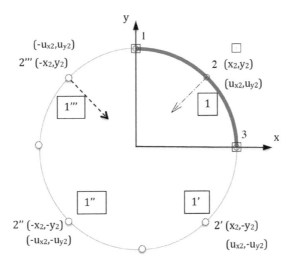

Figure 6 Example of how symmetry conditions are implemented: Original NURBS patch and mirrored patches, showing coordinates and changes in sign for elasticity problems.

We show in Figure 6 that the coordinates of the collocation points and the values of the unknown on mirrored patches can be deduced from patch 1 and therefore the number of unknowns can be reduced and only 3 collocation points need to be considered for the solution.

For this case the implementation of a symmetry capability proceeds as follows:

- The geometrical definition of the mirrored NURBS patche(s) is inherited from the specified NURBS patche(s), except that some coordinates change sign as shown in Figure 6.
- Mirrored patches inherit the incidence and destination vector of the NURBS patch.
- During assembly of the mirrored patches, we have to consider that certain components of **u** have a sign change. This means that some coefficients have to be multiplied by -1 before assembly. For elasticity problems the sign change is shown in Figure 6. For potential problems there is no sign change.
- The outward normal vector for some mirrored patches (in the example they are 1' and 1''') need to be reversed

5   POSTPROCESSING

5.1   Results on the boundary

After the solution we first compute the values of the unknown on the boundary, keeping in mind that the primary results we obtained are parameters not physical values.

We compute $\mathbf{u}(\mathrm{u})$ at a point with the local coordinate u inside a NURBS patch by

$$\mathbf{u}(\mathrm{u}) = \sum_{i=1}^{I} N_i^d(\mathrm{u}) \mathbf{u}_i^e \tag{45}$$

The local parameters are extracted from the global parameters by:

$$\mathbf{u}_i^e = \mathbf{u}_{inci(e,i)} \tag{46}$$

where $inci(e, i)$ is the global number of the i-th parameter of patch e.

Stress recovery   Although we can use derivatives of the fundamental solutions for computing the flow vector and the stress tensor as we will see later, this can not be used directly at the boundary because of the high singularity of the Kernel. Here we propose a simpler, approximate method which is similar to the *stress recovery method* in the FEM [4].

The flow vector in the direction tangential to the boundary $(\bar{x})$ can be computed by:

$$q_{\bar{x}} = -k \frac{\partial u}{\partial \bar{x}} = -k \sum_{i=1}^{I} \frac{\partial N_i}{\partial u} \cdot \frac{\partial u}{\partial \bar{x}} \cdot u_i^e \tag{47}$$

where $\frac{\partial u}{\partial \bar{x}} = J^{-1}$.

For elasticity problems we first work out the strain along a boundary element by

$$\varepsilon_{\bar{x}} = \frac{\partial u_{\bar{x}}}{\partial \bar{x}} = \frac{\partial (\mathbf{u} \cdot \mathbf{v}_1)}{\partial \bar{x}} = \frac{\partial (\mathbf{u} \cdot \mathbf{v}_1)}{\partial u} \frac{\partial u}{\partial \bar{x}} = \frac{\partial (\mathbf{u} \cdot \mathbf{v}_1)}{\partial u} J^{-1} \tag{48}$$

where $\mathbf{v}_1$ is the normalized tangential vector $\mathbf{V}_1$ and

$$\frac{\partial (\mathbf{u} \cdot \mathbf{v}_1)}{\partial u} = \frac{\partial u}{\partial u} \cdot \mathbf{v}_1 + \mathbf{u} \cdot \frac{\partial \mathbf{v}_1}{\partial u} \tag{49}$$

If the Jacobian J is nearly constant along a patch we can use the following approximation:

$$\frac{\partial \mathbf{v}_1}{\partial u} = \frac{\partial^2 \mathbf{x}}{\partial u^2} J^{-1} \tag{50}$$

As this involves a second derivative this term can sometimes be neglected for moderately curved boundary elements.

The stress in tangential direction can then be computed using Hooke's law by

$$\sigma_{\bar{x}} = E \, \varepsilon_{\bar{x}} + \nu \, t_{\bar{y}} + \sigma_{\bar{x}0} \tag{51}$$

where $t_{\bar{y}}$ is the traction normal to the boundary and $\sigma_{\bar{x}0}$ is the tangential component of the initial stress.

5.2   Results inside the domain

Potential/Displacements   We can compute the values of $\mathbf{u}$ at any point $\mathbf{y}$ by using the discretized form of the Somigliana identity:

$$\mathbf{u}(\mathbf{y}) = \sum_{e=1}^{E} \Delta \mathbf{U} \mathbf{t}^e - \sum_{e=1}^{E} \Delta \mathbf{T} \mathbf{u}^e \tag{52}$$

where

$$\triangle \mathbf{U}^e = \int_0^1 \mathbf{U}\big(\mathbf{y}, \mathbf{x}^e(u)\big) \cdot \mathbf{t}(u) J \, du$$

$$\triangle \mathbf{T}^e = \int_0^1 \mathbf{T}\big(\mathbf{y}, \mathbf{x}^e(u)\big) \cdot \mathbf{u}(u) J \, du \tag{53}$$

and for other than Robin boundary condition:

$$\mathbf{u}^e(u) = \sum_{i=1}^{I} N_i(u) \cdot \mathbf{u}_i^e \tag{54}$$

$$\mathbf{t}^e(u) = \sum_{i=1}^{I} N_i(u) \cdot \mathbf{t}_i^e \tag{55}$$

In the case of a Robin BC $\mathbf{t}^e(u)$ is computed using Equation (37).

Flow/Stress   For potential problems the flow vector is computed by

$$\mathbf{q}(\mathbf{y}) = -k \cdot \left(\sum_{e=1}^{E} \triangle \mathbf{U}'^e - \sum_{e=1}^{E} \triangle \mathbf{T}'^e \right) \tag{56}$$

where

$$\triangle \mathbf{U}'^e = \int_0^1 \begin{pmatrix} \dfrac{\partial U}{\partial x} \\[2mm] \dfrac{\partial U}{\partial y} \end{pmatrix} t(u) J \, du$$

$$\triangle \mathbf{T}'^e = \int_0^1 \begin{pmatrix} \dfrac{\partial T}{\partial x} \\[2mm] \dfrac{\partial T}{\partial y} \end{pmatrix} u(u) J \, du \tag{57}$$

For elasticity problems the stresses are collected in a pseudo-vector using the Voight notation and are given by:

$$\boldsymbol{\sigma}(\mathbf{y}) = \sum_{e=1}^{E} \triangle \mathbf{S}^e - \sum_{e=1}^{E} \triangle \mathbf{R}^e \tag{58}$$

where

$$\triangle \mathbf{S}^e = \int_0^1 \mathbf{S} \cdot \mathbf{t}(u) J \, du$$

$$\triangle \mathbf{R}^e = \int_0^1 \mathbf{R} \cdot \mathbf{u}(u) J \, du \tag{59}$$

Table 2 Required number of Gauss points for internal point computation.

R/L	∢2.3187	∢0.9709
M	3	4

The matrices of fundamental solutions **S** and **R** are presented in the Appendix. Since the Kernels are now highly singular i.e. $O(\frac{1}{r^2})$ we use a different criterion for determining the number of Gauss points (see table 2).

6 PROGRAMMING

Here we discuss the implementation of the theory. The function BEM2D shown here is capable of analyzing plane potential and elasticity problems. Indeed, this is one of the nice features of the BEM in that both can be considered in one program with a small amount of additional code.

The input data is split into 5 different files:

- General data, such as analysis type, properties, number of patches, domain type and symmetry code
- Knot vectors describing the geometry of patches
- Control point coordinates and weights describing the geometry of patches
- Boundary conditions
- Refinement data

The last item is necessary to implement the geometry independent approximation approach. It is also convenient for convergence studies. Here it will be assumed that refinement starts from the basis functions used for the geometry definition. This is implemented for convenience, since it saves additional input data, describing the basis functions for the approximation, but can be easily changed to make the program more general.

Input data

The format of the input data is free format and as follows:

File **Input** (general data):

```
Analysis type (1= Potential, 2= Elasticity)
Youngs modulus or conductivity
Poissons ratio (=0 for potential problems)
Number of patches
Domain type (0= finite; 1= infinite)
Symmetry code (1= no symmetry; 2= y-axis symmetry,4= x,y symmetry)
```

File **Knot** (Knot vectors):

```
Number of control points patch 1
Order of patch 1
knot vector of patch 1
Number of control points patch 2
Order of patch 2
knot vector of patch 2
....
```

File **Cntrl** (Control point coordinates and weights):

```
x y z w % coordinates and weight of control point 1 of patch 1
x y z w % coordinates and weight of control point 2 of patch 1
....
x y z w % coordinates and weight of control point 1 of patch 2
x y z w % coordinates and weight of control point 2 of patch 2
....
```

File **BC** (Boundary conditions):

```
BC (1= Dirichlet, 2= Neumann,3=Robin), values  patch no 1
...     %  for all patches
```

File **Ref** (Refinement data):

```
Number of order elevations, Number of knot insertions, knot values
  patch no 1
....     % for all patches
```

Notes:

- When inputting control point coordinates and weights these are defined for each patch separately, so duplicate entries may occur.
- The current implementation is restricted to one BC per patch.
- A $z = 0$ value has to be input.

Description of program BEM2D

In order to keep parameter lists short some global variables are declared at the beginning. These relate to the number of parameters (npa), number of basis functions (nca), region type (rtype), symmetry number (nsy) and number of mirrored patches (nsym).

The program starts with a call to *Readinfo* that reads the input files and makes available the following information to the program: Number of patches (nbs), Knot vectors (Knotg), Coefficients (Coefs), boundary conditions (BC) and Refinement Data (Ref). This information is copied onto the file "Output" that will contain the results. The azimuthal integral is assigned next, which is set to zero for a finite region.

With this information the collocation point coordinates can be computed using function *Colloc*. The function provides the following information: Incidences (inci), local coordinates of collocation points (loco), global coordinates of collocation points (xyp), number of collocation points (nce), indicator if integration is singular (Ising) and information required for the rigid body trick (NBF, RIP, INCIP). Next, function *Destination* computes information required for the assembly. It produces 2 arrays: Ndest and Ldest which specify the row and column number of the global array into which the coefficients are to be assembled.

It also determines an Array Ncode that contains global BCs[1]. The arrays LHS and RHS that will contain the left hand side and the right hand side of the system of equations are set to zero. Array Tn, that will contain the sum of the integrals of **T**, required for the rigid body trick is set to zero. Pointers to information about patches and refinement are also set to zero. In order to check the integration procedure is working correctly we compute the total length and store it in variable Length.

In a loop over all NURBS patches the boundary conditions are assigned via *GetBC*, the geometry information is retrieved via *Get_infoc* and the information about refinement via *Refine*. With this information the structure for describing the geometry (*nurbs*) and for the approximation of the unknown (*nurbsr*) is defined. The number of basis functions for the description of the unknown is *nca*. In a further loop over the symmetry planes *nsym*, mirrored geometry definitions and multiplication factors for assembly (mult) are defined for *nsy > 1*.

In a loop over all collocation points *nptnts* the following is done: *Intscheme* determines if the integration is singular (ising = 1), the number of Gauss points (ngp), their local and global coordinates (Ug, xg), the Jacobian times the weights (JW) and the outward normals (Norm). For the singular integration we have put the number of Gauss points at 4, based on experience.

IntegrateKF preforms the integration and provides arrays dUn, dTn that contain the patch coefficients related to collocation point n. Arrays dRn, dTn contain the value of the basis function and the integral of **T** at collocation point n, required for the rigid body trick. *Store* stores the coefficients in a form suitable for assembly. Finally *Assemb* assembles the coefficient matrices. The coefficients are stored according to the BC and multiplied with a factor before assembly, in the case of symmetry.

Function *Rigidbody* applies the rigid body trick i.e. performs the operation inside the square parentheses in Equation (22). *Plotbv* computes the values at the boundary and returns the values and locations of the maximum and minimum values.

---

[1]In the case of the cantilever for example the Dirichlet BC is valid for the point although it is specified locally for the patches connecting to the point.

Program listing

```
function BEM2D
%--------------------------------------------
% plane Boundary Element program with NURBS
% Programmed by G. Beer 2014
%--------------------------------------------
global npa; global nca; global rtype; global nsy; global nsym
%  Read input data
[nbs,Knotg,Coefs,BC,Ref]= Readinfo;
fout= fopen("Output","w"); Printinfo(fout,nbs,Knotg,Coefs,BC,Ref)
mult(1:npa)=1;  % multiplication factor for symmetry
if(rtype == 1) azi=eye(npa,npa); else azi=zeros(npa,npa); endif
% Compute collocation point coordinates, incidences and destination vectors
[inci,loco,xyp,npnts,nce,Ising,NBF,RIP,INCIP]= Colloc(nbs,Knotg,Coefs,Ref);
[Ndest,Ldest,Ncode,ndofs]= Destination(npnts,xyp,nbs,nce,inci,BC);
%   Compute and assemble coefficient matrices
LHS= zeros(ndofs,ndofs); RHS= zeros(ndofs,1); Tn= zeros(npa,npa,npnts);
i=0; nc=0;nbc=0;Length=0;nr=0;
for nb=1:nbs     % loop over patches
 [bc,values,nbc]= GetBC(nbc,BC);
 [knotu,coefs,i,nc]= Get_infoc(Knotg,Coefs,i,nc);
 nurbs= nrbmak(coefs,knotu);         % definition of geometry
 [nurbsr,nr]= Refine(Ref,nr,nurbs); % approximation of unknown
 knotur= nurbsr.knots; nca= nurbsr.number;   % refined knot vector
 for nsy=1:nsym      % Symmetry loop
  if(nsy > 1)
   [coefss,mult]=Symm(coefs);
   nurbs= nrbmak(coefss,knotu);     % geometry of mirrored patches
  endif
  for npnt=1:npnts         % Loop over collocation points
   xcol(1:2)= xyp(1:2,npnt);
   [ising,ngp,Ug,xg,JW,Norm]= Intscheme(nb,npnt,nurbs,xcol,inci,loco,
   knotur,Ising);
   [dUn,dTn,dRn,Tn,Len]= IntegrateKf(npnt,xcol,ngp,xg,JW,Norm,Ug,nurbsr,Tn,
   bc,values);
   [dU,dT,dR]= Store(Ndest,npnt,dRn,dUn,dTn)  % store patch integrals
  end
  [LHS,RHS]= Assembly(LHS,RHS,nb,bc,dU,dT,dR,values,mult,Ldest,Ncode);
 end
end
for npnt=1:npnts
 Tn(:,:,npnt)= Tn(:,:,npnt) - azi; %   Substract azimuthal integral
end
% Rigid body motion trick
[LHS,RHS]= Rigidbody(npnts,Tn,NBF,RIP,INCIP,LHS,RHS,Ncode);
u= LHS\RHS  % solve
% Get boundary values
nsy=1;
[umax,umin,locmax,locmin,qmax,qmin,locmaxs,locmins]=Plotbv(nbs,u,inci,
Ref,BC,Knotg,Coefs);
endfunction
```

```
function [inci,loco,xyp,npnts,nce,Ising,NBF,RIP,INCIP]= Colloc(nbs,
Knotg,Coefs,Ref)
%-------------------------------------------------------------
% Computes collocation point coordinates and required information.
%
% Input:
% nbs   ...   number of NURBS
% Knotg  ... array with knot vectors
% Coefs  ... array with control point coords and weights
% Ref ...   refinement information
%
% Output:
% inci ...   incidences
% loco ... local coordinates of colloc. points
% xyp  ... global coords of colloc. points
% npts ... number of collocation points
% nce  ... number of parameters for patch
% Ising ... indicator for singular integration
% NBF  ... number of non-zero basis functions at colloc. points
% RIP  ...   basis function values at colloc. points
% INCIP  ...   global numbers of basis functions
%-------------------------------------------------------------
global nsym
npnts= 0;i=0; nc=0;nr=0;
for nb=1:nbs
%---------------
%   Definition of geometry
%---------------
 [knotu,coefs,i,nc]= Get_infoc(Knotg,Coefs,i,nc);
 nurbs= nrbmak(coefs,knotu);
%-------------------
% Approximation of the unknown
%------------------
 nurbsr= Refine(Ref,nr,nurbs); ncr= nurbsr.number;
 pr= nurbsr.order-1;   nce(nb)=ncr;   knotr= nurbsr.knots;
%------------------------
%   Collocation point coordinates
%------------------------
 ut= Greville(ncr,pr,knotr);   xy = nrbeval(nurbs,ut);
 clear nofun; [Rip, nofun] = Nurbbasisfun (ut, nurbsr);
 nfunc= columns(nofun);     npta=npnts;
 for ncu=1:ncr
  xyc(1:2)= xy(1:2,ncu);
  % check if first appearance of point
  if(npnts > 0) [nfirst,npnt]= Check(xyc,xyp,npnts); else nfirst=1; endif
  if(nfirst == 1)
   npnts= npnts+1;   xyp(:,npnts)= xyc(:);
   npnt= npnts;
  endif
  inci(nb,ncu)= npnt;   loco(nb,npnt)= ut(ncu);
 end
```

```
% information for rigid body mode
 for ncu=1:ncr
   npnt= inci(nb,ncu); NBF(npnt)= nfunc;
   for nfu=1:nfunc
     nc= nofun(ncu,nfu); INCIP(nfu,npnt)= inci(nb,nc);
     RIP(nfu,npnt)= Rip(ncu,nfu);
   end
 end
end
%---------------------
%  coll. point indicator, 1= inside patch, 0= otherwise
%--------------------
Ising= zeros(nbs,npnts,nsym);
for nb=1:nbs
 ncr=nce(nb);
 for ncu=1:ncr
  npnt= inci(nb,ncu); Ising(nb,npnt,1)= 1;
  if(nsym == 2)
   if(xyp(1,npnt) == 0) Ising(nb,npnt,2)=1; endif
  else
   if(xyp(2,npnt) == 0) Ising(nb,npnt,2)=1; endif
   if(xyp(1,npnt) == 0) Ising(nb,npnt,4)=1; endif
  endif
 end
end
endfunction
```

```
function [Ndest,Ldest,Ncode,ndofs]= Destination(npnts,xyp,nbs,nce,
inci,BC)
%----------------------------
%  Determines destination vectors
%  Input:
%  npnts ...  no. of coll. points
%  xyp  ...  coords. of coll. points
%  nbs  ...  number of patches
%  nce  ...  no. of patch anchors
%  inci ...  incidences
%  BC  ...  boundary conditions
%
%  Output:
%  Ndest  ...  row destinations
%  Ldest  ...  column destinations
%  Ncode  ...  global BC
%  ndofs  ...  number of degrees of freedom
%----------------------------
global npa; global nsym
%----------------------------
% assign degree of freedom numbers
```

```
%----------------------------
k=0;
for npnt=1:npnts
 for n=1:npa
    k=k+1; Ndest(npnt,n)=k;
 end
end
ndofs=k;
for nb=1:nbs
 for nc=1:nce(nb)
  na= (nc-1)*npa +1 ; ne= nc*npa;
  Ldest(nb,na:ne)= Ndest(inci(nb,nc),1:npa);
 end
end
%------------------
% Global BCs
%------------------
nbc=0; Ncode=zeros(npnts*npa);
for nb=1:nbs
 [bc,values,nbc]= GetBC(nbc,BC);
 for nc=1:nce(nb)
  if(bc == 1)
   npnt=inci(nb,nc);
   for n=1;npa
    ld= Ndest(npnt,n); if(ld > 0) Ncode(ld)=1; endif
   end
  endif
 end
end
endfunction
```

```
function [ising,ngp,Ug,xg,JW,Norm]=
    Intscheme(nb,npnt,nurb,xcol,inci,loco,knotur,Ising)
%----------------------------
% Determines Gauss point locations, Jacobian and outward normal
% for integration for collocation point npnt
%
% Input:
% nb ... number of patch
% npnt ... collocation point
% nurb ...  NURB structure for geometry
% xcol  ...  Collocation point coordinates
% inci  ...  incidences
% loco  ... local coordinate of collocation point
% knotur ...  knot vector for approximation
% Ising ...   indicator if collocation point is in patch
```

```
%
% Output:
% ising  ...  singularity indicator
% ngp    ...   Total number of Gauss points
% Ug     ...   Local coordinates of Gauss points
% xg     ...   Global coordinates of Gauss points
% JW     ...   Jacobian at Gauss points * weights
% Norm   ...   outward normal at Gauss points
%-------------------------------
global nca; global nsy; global nsym
nsing= Ising(nb,npnt,nsy);   ucol= loco(nb,npnt);
[subs,stas] = Subdiv(knotur);   % Determine number of subregions
ngp=0;
for sub=1:subs;               % Loop over subregions
  us(1)= stas(sub); us(2)= stas(sub+1); ising=0;
  if (nsing == 1)
   if(ucol >= us(1) && ucol <= us(2)) ising=1; endif
  endif
  su= us(2)-us(1);
  if(ising == 1)
%--------------------
%   singular integration
%--------------------
    ngaus= 4;   [Cor,Wi]= Gauss(ngaus);
    [nsub,ncase]= Case(ucol,us);
    for ns=1:nsub
     for ng=1:ngaus
      [uga,Jacga]= Transform(ns,nsub,ncase,su,Cor(ng),us);
      ug(ng)=uga; Jacg(ng)=Jacga;
     end
     [xy,jac,norm]=Diffgeom(nurb,ug);
     for ng=1:ngaus
      ngp=ngp+1; xg(:,ngp)= xy(:,ng);
      JW(ngp)= jac(ng)*Jacg(ng)*Wi(ng);
      Norm(:,ngp)= norm(:,ng); Ug(ngp)= ug(ng);
     end
    end
   else
%--------------------
% regular integration
%--------------------
    [Rmin,L]= Mindist(nurb,xcol,us,su);
    [ngaus,ndivs]= Gausspoints(Rmin,L,1);
    [Cor,Wi]= Gauss(ngaus); xsi1=-1; divs=single(ndivs);
    xdelt= 2/divs;
    for ndiv=1:ndivs
     xsi2=xsi1 + xdelt;
     for ng=1:ngaus
      if(ndivs > 1)
       xsi= 0.5*(xsi1 + xsi2) + Cor(ng)/divs; Jacx= 1/divs;
      else
```

```
     xsi= Cor(ng); Jacx=1;
    endif
    ug(ng)= su/2*(1+xsi)+us(1);
   end
   Jacu= su/2*Jacx; [xy,jac,norm]=Diffgeom(nurb,ug);
   for ng=1:ngaus
     ngp=ngp+1; xg(:,ngp)= xy(:,ng);
     JW(ngp)= jac(ng)*Jacu*Wi(ng);
     Norm(:,ngp)= norm(:,ng); Ug(ngp)= ug(ng);
   end
   xsi1= xsi2;
  end
 endif
end
endfunction
```

```
function [dUn,dTn,dRn,Tn,Len]=
    IntegrateKf(npnt,xp,ngp,xg,JW,Norm,Ug,nurbsr,Tn,bc,values)
%--------------------------------
% Integrates Kernel-function products
%
% Input:
% npnt ...  collocation point number
% xp  ... collocation point coords
% ngp ... total number of Gauss points
% xg  ...  Gauss point coordinates
% JW ...  Jacobian at Gauss points*weights
% Norm ... outward normal
% Ug ...  local Gauss coords
% nurbsr ... structure for refinement
% Tn ... value of Tn for summation
% bc ... boundary code
% values ... boundary values
%
% Output:
% dU, dTn ... Delta Un or Delta Tn
% dR  ... Delta Rn
% dTn ... Delta Tn
% Len ...  contribution to length
%--------------------------
global npa; global nca
dUn=zeros(npa,npa,nca); dTn=zeros(npa,npa,nca);
dRn= zeros(npa,1); dTsum= zeros(npa,npa); Len=0;
clear nofun; [Ri, nofun] = Nurbbasisfun (Ug, nurbsr);
for ng=1:ngp           % loop over Gauss points
 JWng=JW(ng); Len= Len + JWng;
 norm(1:2)= Norm(1:2,ng); xgaus(1:2)=xg(1:2,ng); r= Dist(xp,xgaus);
 dxr= (xgaus-xp)/r;  UK= UKernel(r,dxr); TK= TKernel(r,dxr,norm);
 dTsum= dTsum + TK*JWng;
```

```
if(bc > 1)
  if(bc == 3)  t0= Robin(norm,values); else t0=values; endif
  dRn(:,1)= dRn(:,1) + UK*t0*JWng;
endif
for j=1:columns(nofun)  %  loop over pointers to basis functions
  nf= nofun(ng,j);   dUn(:,:,nf)= dUn(:,:,nf) + UK*Ri(ng,j)*JWng;
  dTn(:,:,nf)= dTn(:,:,nf) + TK*Ri(ng,j)*JWng;
 end
end
Tn(:,:,npnt)= Tn(:,:,npnt) + dTsum(:,:);
endfunction
```

```
function [dU,dT,dR]= Store(Ndest,npnt,dRn,dUn,dTn)
%----------------------------
%  store patch integrals
%
%  Input:
%  Ndest   ...  row destination
%  npnt  ...  collocation point
%  dRn,dUn,dTn ... Integrals for point nc
%
%  Output:
%  dU,dT,dR  ...  integrals for patch
%---------------
global npa; global nca
for i=1:npa
    nrow= Ndest(npnt,i); if(nrow == 0) continue endif
    dR(nrow,1)= dRn(i,1);
    for nc=1:nca
     ncol= (nc-1)*npa;
     for j=1:npa
      ncol= ncol+1; dU(nrow,ncol)= dUn(i,j,nc);
      dT(nrow,ncol) = dTn(i,j,nc);
     end
    end
end
endfunction
```

```
function [LHS,RHS]= Assembly(LHS,RHS,nb,bc,dU,dT,dR,values,
Mult,Ldest,Ncode)
%----------------------------------------
% Assembly
%-------------------------------
global npa; global nca; global Bcode; global Values
```

```
i=0;
if(bc > 1)    % Robin or Neumann BC
     RHS(:,1)= RHS(:,1) + dR(:,1);
endif
for nc=1:nca
 for n=1:npa
  i=i+1;mult= Mult(n);
  ncol= Ldest(nb,i);
  if(ncol == 0 ) continue endif
  if(bc > 1)    %  Neumann or Robin BC
    LHS(:,ncol)= LHS(:,ncol) + dT(:,i)*mult;
  else          %  Dirichlet BC
    LHS(:,ncol)= LHS(:,ncol) - dU(:,i)*mult;
    RHS(:,1)= RHS(:,1) - dt(:,i)*values*mult;
  endif
 end
end
endfunction
```

```
function [LHS,RHS]= Rigidbody(npnts,Tn,NBF,RIP,INCIP,LHS,RHS,Ncode)
%----------------------------------------
% Apply rigid body motion
%
% Input:
% npnts  ...  number of coll. points
% Tn ... sum of T-terms
% NBF  ... number of non-zero basis functions
% RIP  ...  basis function values
% INCIP  ...  global numbers of basis functions
% LHS,RHS  ...  left and right hand side
% Ncode ... global BC
%
% Output:
% LHS,RHS  ...  updated left and right hand side
%-------------------------------
global npa; global Values
for npt=1:npnts
 if(Ncode(npt) == 1 ) continue endif  % Dirichlet BC
 nf=NBF(npt);
 for nfu=1:nf
  RI=RIP(nfu,npt);
  if(RI > 0)
    IP= INCIP(nfu,npt); nr= (npt-1)*2;
    for i=1:2
     nr=nr+1; nc=(IP-1)*2;
```

```
   for j=1:2
    nc= nc+1; LHS(nr,nc)= LHS(nr,nc) - Tn(i,j,npt)*RI;
   end
  end
 endif
 end
end
endfunction
```

7 EXAMPLES

7.1 Potential problem: Flow past isolator

We revisit the problem of the flow past an isolator that we introduced previously. We solve two problems: One without the hole and the flow applied and one where the negative flow normal to the boundary is applied. The second is a Robin BC problem with the BC computed by:

$$t_0 = \mathbf{q}_0 \cdot \mathbf{n} \tag{60}$$

Here we assume $\mathbf{q}_0 = \begin{pmatrix} 0 \\ 1 \end{pmatrix}$, symmetry and only discretize half of the problem with one NURBS patch.

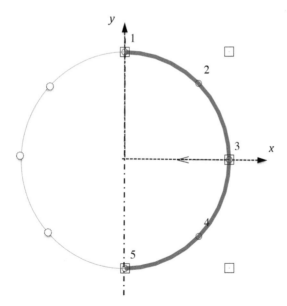

Figure 7 Discretization of half the boundary with one NURBS patch, showing control points (squares) and collocation points (circles).

The input data for this problem are:

File **Input:**

```
1
1.0
0
1
1
2
```

File **Knot:**

```
5
2
0 0 0  0.5 0.5 1 1 1
```

File **Cntrl:**

```
0  1  0  1
1  1  0  0.707
1  0  0  1
1 -1  0  0.707
0 -1  0  1
```

File **BC:**

```
3  0  -1
```

File **Ref:**

```
0  0
```

The maximum Temperature obtained from the program is 2.0004 as compared with an exact value of 2.0. In this case the geometry, the distribution of the known boundary values and the variation of the solution are all exactly represented. Therefore, no refinement of the solution is necessary. The small error is only due to the precision of the numerical integration and the indication is that the integration criterion proposed earlier is adequate.

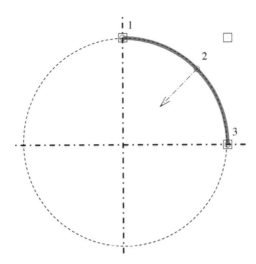

Figure 8 Discretisation of circular excavation with 1 patch and 2 axes of symmetry. Control points depicted by squares and collocation points by circles.

This is a good example of the power of the NURBS based simulation: an exact definition of the geometry, the known boundary values and in this case even the variation of the unknown.

7.2   Elasticity problem: Circular excavation in infinite domain

This is similar problem to the one discussed in stage 4 (infinite plate with a hole) except that the extent of the domain is really infinite. Since there is no finite boundary on which to apply the Neuman BC's the problem is solved in a similar way to the previous one.

We solve two problems: One without the hole and the virgin stress σ_{0x} and σ_{0y} applied and one where the excavation tractions are applied. The second is a Robin BC problem with the tractions computed by:

$$\mathbf{t}_0 = \begin{pmatrix} n_x \sigma_{0x} \\ n_y \sigma_{0y} \end{pmatrix} \tag{61}$$

We assume a virgin stress field of $(\sigma_{x0} = 1, \sigma_{y0} = -0)$, two planes of symmetry and a discretization with one NURBS patch as shown in Figure 8.

The input data for this problem are:

File **Input:**

```
1    1.0  0  1  1  2
```

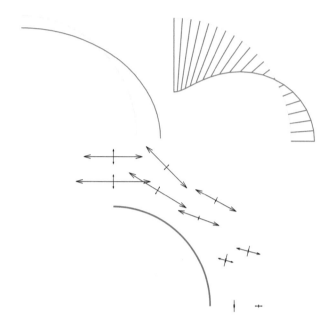

Figure 9 Displaced shape of 1/4 circle, distribution stress in direction tangential to the boundary and principal stress vectors.

File **Knot:**

```
3   2
0  0  0  1  1  1
```

File **Cntrl:**

```
0  1  0  1
1  1  0  0.707
1  0  0  1
```

File **BC:**

```
3  1  0
```

File **Ref:**

```
0  0
```

The maximum x-displacement computed is 2.005 compared with the theoretical value of 2.0. The maximum tangential stress, computed by stress recovery is 3.002 as compared with the exact value of 3.0.

As in this case the geometry, the variation of the known values and the variation of the unknown is exactly represented, the error is only due to the integration error. The computed displaced shape, the variation of the tangential stress along the boundary and principal stress vectors are shown in Figure 9.

7.3 Practical example: Horseshoe tunnel

The geometry description of this practical application was already presented previously and is an exact representation of the design geometry. Here we simulate the excavation of the tunnel in a domain subjected to a virgin stress field of $(\sigma_{x0} = 0, \sigma_{y0} = -1)$.

The input data are:

File **Input:**

```
2 1.0 0.0 1 1 2
```

File **Knot:**

```
7   2
0 0 0 0.5 0.5 0.8 0.8 1 1 1
```

File **Cntrl:**

```
0.0     5.65   0 1.0
4.55 5.65     0 0.707
4.55 1.1      0 1.0
4.55 -0.97 0 0.82
2.61 -1.67 0 1.0
1.33 -2.04 0   0.99
0.0 -2.04    0 1.0
```

File **BC:**

```
3 0 -1
```

File **Ref** pior to refinement:

```
0 0
```

Two refinement strategies were applied. One was to insert knots into the knot vectors describing the geometry and the oder was a k-refinement, i.e. the order of the basis functions was increased before knot insertion. As a measure of convergence we

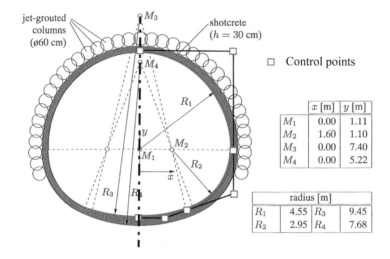

	x [m]	y [m]
M_1	0.00	1.11
M_2	1.60	1.10
M_3	0.00	7.40
M_4	0.00	5.22

radius [m]			
R_1	4.55	R_3	9.45
R_2	2.95	R_4	7.68

Figure 10 Definition of the geometry of a horseshoe tunnel showing control points.

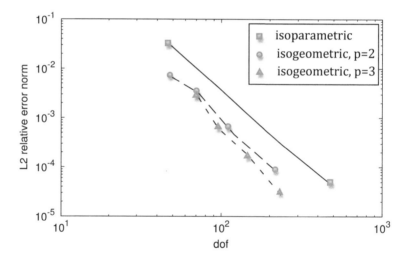

Figure 11 Convergence of the solution for the isoparametric and NURBS based BEM.

use the L2 norm that was introduced in stage 4. A comparison with the conventional BEM is also included. A plot of the L2 norm versus the degrees of freedom (dof) (Figure 11) shows that the NURBS based BEM requires fewer degrees of freedom and converges faster than the isoparametric BEM. The converged results of the simulation are shown in Figure 12.

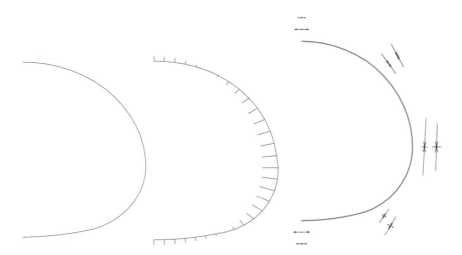

Figure 12   Displaced shape of half the tunnel, distribution stress in direction tangential to the boundary and principal stress vectors.

8   CONCLUSIONS

Here we have introduced the BEM, which has been far less popular than its big sister the FEM. The main reason for its introduction here was that it is an ideal companion of CAD, since both rely on a boundary definition. We argue that the goal of a seamless integration of CAD and simulation would only be possible via the BEM.

It is hoped that readers could appreciate the beauty of the method: The problem definition has been reduced by one order (surface instead of volume definition) and the solutions actually satisfy the governing differential equations and are therefore more accurate than with the FEM.

On the examples of the circular isolator and excavation it was shown that the exact solution can be obtained with a very coarse mesh, because the geometry and the unknown can be defined exactly by a NURBS of order 2.

It is acknowledged that due to the need of dealing with singular integrals, the implementation is more involved than with the FEM, but this is a small price to pay considering the quality of the results, the user friendliness of the method and the potential it has for combining it with CAD. At the next stage, the implementation of the NURBS based BEM in 3-D is discussed.

BIBLIOGRAPHY

[1]   U. Eberwien, C. Duenser, and W. Moser. Efficient calculation of internal results in 2D elasticity BEM. *Engineering Analysis with Boundary Elements*, 29(5):447–453, 2005.

[2]   J.C. Lachat and J.O. Watson. Effective numerical treatment of boundary integral equations: A formulation for three-dimensional elastostatics. *International Journal for Numerical Methods in Engineering*, 10(5):991–1005, 1976.

[3]   Peter R. Johnston and David Elliott. A generalisation of Telles' method for evaluating weakly singular boundary element integrals. *Journal of Computational and Applied Mathematics*, 131(1–2):223–241, June 2001.

[4]   I. M. Smith, D. V. Griffiths, and L. Margetts. *Programming the Finite Element Method*. Wiley, 2013.

[5]   A.H. Stroud and D. Secrest. *Gaussian Quadrature Formulas*. Prentice Hall, Englewood Cliffs, New Jersey, 1966.

[6]   J.O. Watson. *Developments in Boundary Element Methods – 1*, volume 1, chapter Advanced implementation of the boundary element method for two- and threedimensional elastostatics. Applied Science, 1979.

Chapter 9

Stage 8: The boundary element method for three-dimensional problems

where we expand into 3-D space.

1   INTRODUCTION

The extension of the method into 3-D space follows a similar approach as for the plane case. Here we start immediately with the NURBS based approach, without going through the classical approach first.

Discretization   We start with the definition of the geometry and the approximation of boundary values:

$$\mathbf{x}^e = \sum_{k=1}^{K} R_k(\mathrm{u}, \mathrm{v}) \cdot \mathbf{x}_k^e \tag{1}$$

$$\mathbf{u}^e = \sum_{k=1}^{K^d} R_k^d(\mathrm{u}, \mathrm{v}) \cdot \mathbf{u}_k^e \tag{2}$$

$$\mathbf{t}^e = \sum_{k=1}^{K^t} R_k^t(\mathrm{u}, \mathrm{v}) \cdot \mathbf{t}_k^e \tag{3}$$

where R_k, R_k^d, R_k^t are suitable basis functions of the local coordinate u, v for describing the geometry, displacements and tractions, $\mathbf{x}_k^e$ specify the location of control points, $\mathbf{u}_k^e$, $\mathbf{t}_k^e$ are parameter values of $\mathbf{u}$, $\mathbf{t}$ and K, K^d, K^t are the total number of parameters.

Collocation points   The first task is to compute the location of the collocation points. As explained previously the location of the points is equal to the anchors of basis functions used to approximate the solution. The local coordinates of the nth collocation point are given by:

$$\begin{aligned}
\mathrm{u}_{i(n)} &= \frac{\mathrm{u}_{i+1} + \mathrm{u}_{i+2} + \cdots + \mathrm{u}_{i+p_u}}{p_u} \quad i = 0, 1, \ldots, I \\[2mm]
\mathrm{v}_{j(n)} &= \frac{\mathrm{v}_{j+1} + \mathrm{v}_{j+2} + \cdots + \mathrm{v}_{j+p_v}}{p_v} \quad j = 0, 1, \ldots, J
\end{aligned} \tag{4}$$

where $i(n)$ and $j(n)$ specify local numbering of collocation point n in u and v directions, p_u, p_v are the orders and I and J specifies the number of control points in u and v directions.

2　NUMERICAL INTEGRATION

Expanding the discretized integral equation to 3D:

$$\sum_{e=1}^{E}\sum_{k=1}^{K^d}\Delta T_{ni}^e u_i^e - \left[\sum_{k=1}^{K^d}R_k^d(u_n,v_n)u_k^{en}\right]\sum_{e=1}^{E}T_n^e = \sum_{e=1}^{E}\sum_{k=1}^{K^t}\Delta U_{ni}^e t_k^e \tag{5}$$

where ΔT_{ni}^e and ΔU_{ni}^e are integrals of Kernel basis function products and T_n^e is an integral of the T Kernel only. As for the plane case we distinguish between regular and weakly singular integrals.

2.1　Regular integration

We recall that the region of integration may have to be subdivided into sub-regions in the cases where some basis functions have limited span. To use Gauss Quadrature the sub-region must be mapped into a coordinate system ξ, η that ranges from -1 to 1 by

$$u = \frac{s_u}{2}(\xi + 1) + u_s \tag{6}$$

$$v = \frac{s_v}{2}(\eta + 1) + v_s$$

where u_s, v_s denotes the local coordinates of the start and s_u and s_v denote the size of the subregion. The Jacobian of this transformation is

$$J_s = \frac{du}{d\xi}\cdot\frac{dv}{d\eta} = \frac{s_u}{2}\cdot\frac{s_v}{2} \tag{7}$$

The regular integrals are given by

$$\Delta U_{n,k}^e = \sum_{s=1}^{S}\int_{-1}^{+1}\int_{-1}^{+1} U(\mathbf{y}_n,\mathbf{x})R_k^t JJ_s\,d\xi\,d\eta$$

$$= \sum_{s=1}^{S}\sum_{m=1}^{M}\sum_{l=1}^{L}U(\mathbf{y}_n,\mathbf{x}(u_m,v_l))R_k^t(u_m,v_l)JJ_s\,W_m W_l$$

$$\Delta T_{n,k}^e = \sum_{s=1}^{S}\int_{-1}^{+1}\int_{-1}^{+1} T(\mathbf{y}_n,\mathbf{x})R_k^d JJ_s\,d\xi\,d\eta \tag{8}$$

$$= \sum_{s=1}^{S}\sum_{m=1}^{M}\sum_{l=1}^{L}T(\mathbf{y}_n,\mathbf{x}(u_m,v_l))R_k^d(u_m,v_l)JJ_s\,W_m W_l$$

$$\Delta T_n^e = \sum_{s=1}^{S}\int_{-1}^{+1}\int_{-1}^{+1} T(\mathbf{y}_n,\mathbf{x})JJ_s\,d\xi\,d\eta = \sum_{s=1}^{S}\sum_{m=1}^{M}\sum_{l=1}^{L}T(\mathbf{y}_n,\mathbf{x}(u_m,v_l))JJ_s\,W_m W_l$$

In the above M, L are the number of Gauss points in ξ, η direction (chosen depending on the proximity of the collocation point to the element), $u_m(\xi)$, $v_l(\eta)$ are coordinates of Gauss points and W_m, W_l are weights. The sum is over the number of sub-regions S.

The choice of the right number of Gauss points, depending on the proximity of the collocation point is now crucial because the number of operations per Gauss point is larger than for the plane problem and the number of Gauss points increases with the square. The formula to determine the required number of Gauss points, which was derived for the plane case, only based on the order of the Kernel, will no longer suffice.

2.2 Determination of the optimal number of Gauss points

The number of numerical operations to be carried out at each Gauss point is quite high. For the determination of the outward normal and the Jacobian, derivatives of the basis functions and several multiplications are required. The evaluation of the Kernel adds further floating point operations. In order to minimize the computing cost it is desirable to arrive at a minimum number of integration points over a NURBS patch to achieve a given accuracy. For the case where the collocation points are far away from the NURBS patch (far field) the integration rule can be bundled for groups of collocation points so that evaluations at Gauss points are reduced. There is a tradeoff between integration over the whole NURBS patch without subdivision and subdividing the patch into smaller integration elements.

In our endeavor to obtain an optimal number of Gauss points for the case where no subdivision is necessary – due to the limited span of the basis functions – for the approximation of the unknown, we adopt the following philosophy:

- Try to integrate over the whole NURBS patch with one integration region only
- Limit the number of Gauss points to 8
- Subdivide if, according to the rules developed later, this number of Gauss points is not sufficient

The method we propose for obtaining the optimal number of Gauss points is empirical. Experience shows that it is convenient to have only one integration scheme for both the U and the T Kernels. Since the T Kernel has the higher singularity we concentrate on the integration involving this Kernel. We investigate the number of Gauss points required for a given accuracy, depending on the ratio of minimum distance to the collocation point (R) to the length of the integration region (L), $\frac{R}{L}$, for a number of possible locations of the collocation point.

First we consider an integrand without the Kernel. This would correspond to the case where the collocation point is far away from the integration region (in the far field) so that the variation of the Kernel is nearly constant. We investigate the integration of:

$$Int = R_i^d(\mathrm{u}, \mathrm{v}) \cdot J \cdot J_s \tag{9}$$

and how many Gauss points are required for a given accuracy.

We define the integration error as:

$$|\epsilon(M)| = \frac{|Int(M) - Int^{exact}|}{|Int^{exact}|} \tag{10}$$

where Int^{exact} is a value of the integral obtained with a large number of Gauss points (4×4 subdivision of the patch with 8×8 Gauss points each $= 1024$ Gauss points).

Next we consider the whole integrand including the Kernel:

$$\mathbf{Int} = \mathbf{T}\left(\mathbf{y}_n, \mathbf{x}\right) \cdot R_k^d(u, v) \cdot J \cdot J_s \tag{11}$$

Since **Int** is a matrix we define a norm of the difference in the computed value and an *exact* value (that has been computed as explained before) as:

$$\|\epsilon(M)\| = \frac{\|\Delta\mathbf{Int}\|}{\|\mathbf{Int}\|} \tag{12}$$

where

$$\|\Delta\mathbf{Int}\| = \sqrt{\sum_i \sum_j (Int_{ij}^M - Int_{ij}^{exact})^2} \tag{13}$$

$$\|\mathbf{Int}\| = \sqrt{\sum_i \sum_j (Int_{ij}^{exact})^2} \tag{14}$$

Int_{ij}^M is the coefficient i, j of **Int** computed with M integration points and Int_{ij}^{exact} is the *exact* value. We investigate the influence of the proximity of the collocation point and refer to this as near field integration.

2.3 Regular integration

Here we investigate the integration error for the case where the collocation point is far away so that the variation of the Kernel is nearly constant over the NURBS patch.

We consider the integrand without the Kernel and first a fairly regular NURBS patch describing a 1/4 cylinder shown in Figure 1. We quite arbitrarily select the basis function to be the second one (similar results were obtained for the other basis functions) and plot the error as a function of order. First we start with quadratic basis functions (meaning that $p_u = p_v = 2$) followed by elevation of the order 1, 2, 3 times in both directions. The result of this investigation is shown in Figure 3 and shows that unless the order elevation is extreme there is a small influence on the error. Figure 3 can be used for determining the number of Gauss points in each direction for far field integration of fairly regular patches depending on the required maximum error. Next we consider irregular NURBS patches. To test the robustness of the integration we investigate its application to trimmed NURBS.

Two cases are considered: one where the trimming curve is of order 2 and the trimming is moderate and one where the trimming curve is of order 3 and the trimming is severe. The first case is depicted in Figure 4. The results for the far field integration are depicted in Figure 5. It can be seen that for moderate trimming the plot of the

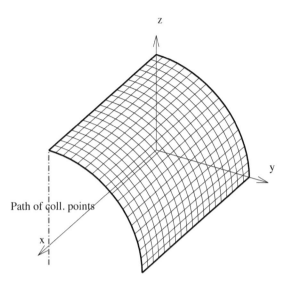

Figure 1 NURBS patch considered for determining the number of Gauss points. Dotted line shows path of collocation point for near field integration.

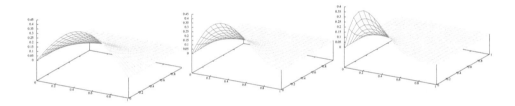

Figure 2 Plot of second basis function for different degrees of order elevation.

number of Gauss points versus error is similar to the plots for the untrimmed surface. For test 2 we apply a trimming curve of order 3 (Figure 6). The results for the far field integration are shown in Figure 7. It shows oscillatory behavior, meaning that this patch can not be integrated without subdivision.

2.4 Nearly singular integration

Next we investigate the near field integration by considering the whole integrand including the Kernel and by placing the collocation point in a vertical line nearer and nearer to the NURBS patch as shown in Figure 1.

In Figure 8 we show the results for the second basis function of order $p_u = p_v = 2$. The graph can be used for the integration strategies discussed next. For an integration error of 10^{-3} for example we find that we need 4 Gauss points in the direction where the length L has been computed, if R/L is greater than 1.0, 5 Gauss points if R/L is

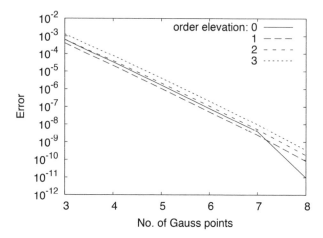

Figure 3 Far field integration: Plot of Number of Gauss points vs. integration error for different order elevations (0 means $p_u = p_v = 2$).

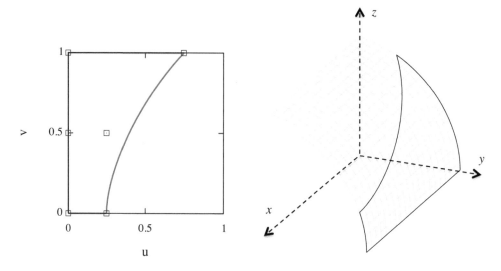

Figure 4 Test 1 of trimmed NURBS: left trimming curves in u, v coordinate system, right trimmed surface.

less than 1.0 and greater than 0.6 and 6 Gauss points if R/L is less than 0.6 and greater than 0.4. For R/L less than 0.4 a subdivision is needed. Table 1 shows the number of Gauss points, M, required depending on the value of R/L not being less than a certain value. A subdivision is required for values smaller than 0.2 for and an error of 10^{-2} and smaller than 0.4 for errors of 10^{-3} and 10^{-4}.

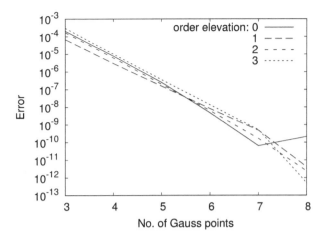

Figure 5 Far field integration of trimmed surface: Plot of number of Gauss points versus order for test 1.

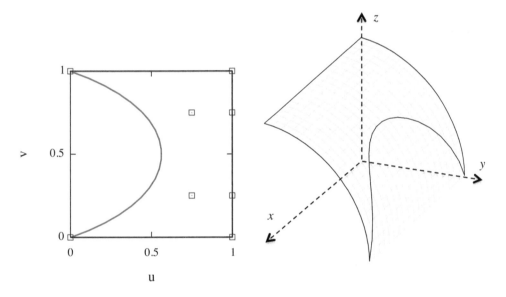

Figure 6 Test 2 of trimmed NURBS: left trimming curves in u, v coordinate system, right trimmed surface.

Subdivision strategies As has become apparent, there will be a need for subdividing the NURBS patch into integration elements. There are basically two strategies for this:

1 Subdivide the patch into as many equal integration regions as required so that each subdivision satisfies that $R/L < Rlim$ where $Rlim$ is a user defined limiting value. We refer to this as the regular subdivision method.

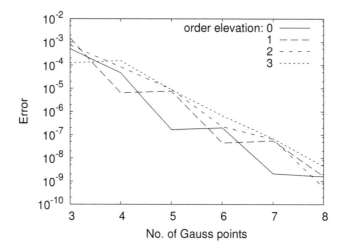

Figure 7 Far field integration of trimmed surface: Plot of number of Gauss points vs order for test 2.

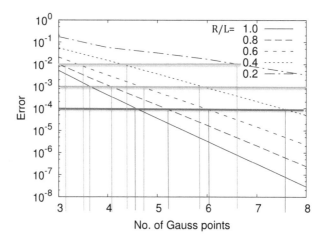

Figure 8 Plot of error versus number of Gauss points for different values R/L. Also shown is how the number of Gauss points can be determined for errors of 10^{-2}, 10^{-3} and 10^{-4}.

Table I Required number of Gauss points depending on the maximum error and value of R/L not being less than a certain value.

Error	$M =$	3	4	5	6	7	8
10^{-2}	$R/L \nless$	0.8	0.6	0.4	0.2		
10^{-3}	$R/L \nless$	1.5	1.0	0.6	0.4		
10^{-4}	$R/L \nless$	2	1.5	1.0	0.8	0.6	0.4

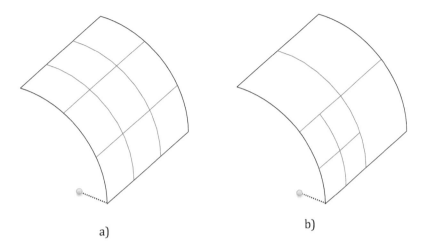

a)   b)

Figure 9 The different strategies for subdivision into integration regions: a) equal subdivision,
b) Quadtree subdividision.

2   Subdivide the patch in 4 equal subelements if $R/L > Rlim$. If the criterion
$R/L < Rlim$ is violated for any integration region subdivide this region further
until the requirement is satisfied. We refer to this as the Quadtree method.

To test the two strategies we place the collocation point very near the integration
region in this case $R/L = 0.15$. For this case a subdivision into integration regions is
needed. We first follow strategy 1, i.e. we subdivide equally in all directions (to satisfy
the integration criterion for an error of 10^{-3} we need 3 subdivisions in each direction).
For each sub region we use the criterion to determine the number of Gauss points
depending on the R/L value computed for the subregion (see Figure 9a). This strategy
results in a total number of Gauss points of 206.

Next we apply strategy 2 and this means that we halve the dimensions of the subregions until the criterion for the minimum value of R/L is satisfied. This is shown in
Figure 9b. This strategy results in a reduction of the total number of Gauss points to
139, i.e. a 33% reduction. The accuracy of the integration for both schemes is roughly
the same but the error has actually decreased to 10^{-4}, although the integration criterion
for 10^{-3} was applied.

2.5 Weakly singular integration

The kernel **U** has a singularity of $O(\frac{1}{r})$. The product $\mathbf{U}R_k$ tends to infinity if R_k does
not tend to zero at the singularity point. For this case we use a method that has been
found to work well for the iso-parametric BEM, i.e. we perform the integration in a
local coordinate system, where the Jacobian tends to zero as the singularity point is
approached. For this we divide the integration region into two, three or four triangular
sub-regions depending if the collocation point is at a corner, edge or inside as shown
in Figure 10.

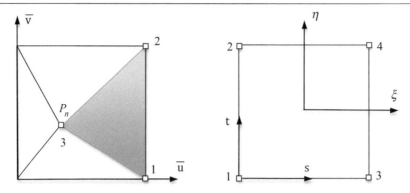

Figure 10 Coordinate transformation between $\bar{u}, \bar{v}, s, t$ and ξ, η systems for the case where the collocation point is inside the domain and for the highlighted triangle. Control points are depicted by rectangles and numbered in the $\bar{u}, \bar{v}$ and s, t coordinate system.

For the transformation between the coordinate systems we can use the mapping introduced earlier for trimmed surfaces. For this we define a NURBS patch of order 1 in an s, t coordinate system with the knot vectors $\Xi_s = \Xi_t = 0, 0, 1, 1$. We denote with $\bar{u}, \bar{v}$ the coordinate system of the subregion of integration and the transformation to NURBS coordinates u,v is given by:

$$u = s_u \cdot \bar{u} + u_s$$

$$v = s_v \cdot \bar{v} + v_s \tag{15}$$

where s_u, s_v specify the size of the subregion and u_s, v_s the starting coordinates in u and v directions. The Jacobian of this transformation is

$$J_s = s_u \cdot s_v \tag{16}$$

The transformation between $\bar{u}, \bar{v}$ and s, t systems is given by

$$\bar{u} = \sum_{i=1}^{4} \bar{R}_i(s, t)\bar{u}_{n(i)}$$

$$\bar{v} = \sum_{i=1}^{4} \bar{R}_i(s, t)\bar{v}_{n(i)} \tag{17}$$

where $\bar{R}_i(s, t)$ are the NURBS basis functions of s,t, $\bar{u}_{n(i)}, \bar{v}_{n(i)}$ are control point coordinates and $n(i)$ denotes the numbering of the point i in the $\bar{u}, \bar{v}$ coordinate system. For the highlighted triangle in Figure 10 we have:

$$n(i) = (3 \quad 3 \quad 1 \quad 2) \tag{18}$$

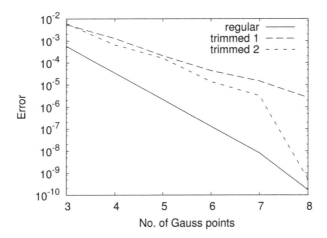

Figure 11 Plot of integration error versus number of Gauss points for weakly singular integration applied to regular and trimmed patches.

The Jacobian of this transformation is J_{tr} and tends to zero as the collocation point is approached. Finally the transformation between the s, t and ξ, η is given by

$$s = \frac{1}{2}(\xi + 1) \tag{19}$$

$$t = \frac{1}{2}(\eta + 1) \tag{20}$$

where the Jacobian is 0.25.

Applying Gauss quadrature we obtain

$$\triangle U_{n,i}^{e} = \sum_{s=1}^{S}\sum_{nt=1}^{ntr}\int_{-1}^{+1}\int_{-1}^{+1} U(P_n, Q)R_k(\mathrm{u}, \mathrm{v}) \cdot J \cdot J_s \cdot J_{tr} \cdot 0.25 \cdot d\xi\,d\eta \tag{21}$$

$$= \sum_{s=1}^{S}\sum_{nt=1}^{ntr}\sum_{m=1}^{M}\sum_{l=1}^{L} U(P_n, Q)R_k(\mathrm{u}(\xi_m), \mathrm{v}(\eta_l)) \cdot J \cdot J_s \cdot J_{tr} \cdot 0.25 \cdot W_m W_l \tag{22}$$

where S is the number of subregions and *ntr* is the number of triangles.

We test the integration accuracy for the 3 cases introduced above, i.e. for a regular patch and for 2 trimmed patches. We assume that the collocation point is a corner node. The result is shown in Figure 11. It can be seen that for weakly singular integration fewer Gauss points are needed as compared with the regular integration. However, this changes when patches are trimmed and more Gauss points are required for trimmed surfaces in this case.

2.6 Infinite patches

The geometry of mapped infinite patches have been discussed previously. Two assumptions of the variation of **u** may be considered.

- *Plane Strain*: Displacements are constant to infinity[1].

$$\mathbf{u}(u, v) = \mathbf{u}(u, v = 0) = \sum_{k=1}^{K^d} R_k^f(u)\mathbf{u}_k^e \tag{23}$$

- *Decay*: Displacements decay to zero as infinity is approached[2].

$$\mathbf{u}(u, v) = (1 - v)\mathbf{u}(u, v = 0) = (1 - v) \sum_{k=1}^{K^d} R_k^f(u)\mathbf{u}_k^e \tag{24}$$

In the above K^d is the number of parameters and $\mathbf{u}_k^e$ are parameter values at the control points at the finite edge of the NURBS patch. $R_k^f(u)$ are basis functions of u describing the variation of **u** on the finite edge. It can be shown (see [6]) that the second assumption translates in a decay of $O(\frac{1}{r})$, which is the appropriate decay for displacements in 3-D elasticity.

The integrals to be evaluated are for a *plane strain* element

$$\int_{S_e} \mathbf{U}(P_n, Q)\mathbf{t}\, dS = \int_{-1}^{+1} \int_{-1}^{+1} \mathbf{U}(P_n, Q)\mathbf{t}\, J\, 0.25\, d\xi\, d\eta$$

$$\int_{S_e} \mathbf{T}(P_n, Q)R_k^f\, dS = \int_{-1}^{+1} \int_{-1}^{+1} \mathbf{T}(P_n, Q)R_k^f\, J\, 0.25\, d\xi\, d\eta \tag{25}$$

and for a *decay* element[3].

$$\int_{S_e} \mathbf{T}(\mathbf{y}_n, \mathbf{x})(1 - v)R_k^f\, dS = \int_{-1}^{+1} \int_{-1}^{+1} \mathbf{T}(\mathbf{y}_n, \mathbf{x})(1 - v)R_k^f\, J\, 0.25\, d\xi\, d\eta \tag{26}$$

For a *plane strain* infinite patch, the integral involving the Kernel **U** varies as $O(\frac{1}{r})$ and because the sides to infinity of the patch must be parallel the Jacobian will vary with $O(r^2)$. Therefore the integral does not have a finite value. It can be shown (see [7]), however, that the integral over a closed contour[4] has a finite value.

---

[1] For example a very long tunnel, where plane strain conditions are attained away from the tunnel face.
[2] For example a surface that extends to infinity.
[3] Here we assume that the tractions are zero over an infinite patch, so the integral with Kernel **U** does not need to be evaluated.
[4] Note that all applications of the plane strain patch must involve a closed contour.

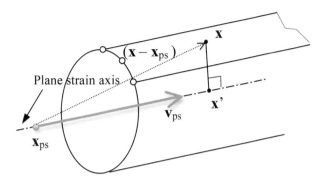

Figure 12 Explanation of the computation of $\mathbf{x}'$.

The integral can be changed to

$$\int_{-1}^{+1} \int_{-1}^{+1} \mathbf{U}(\mathbf{y}_n, \mathbf{x})\mathbf{t}\, 0.25\, d\xi\, d\eta = \int_{-1}^{+1} \int_{-1}^{+1} (\mathbf{U}(\mathbf{y}_n, \mathbf{x}) - \mathbf{U}(\mathbf{y}_n, \mathbf{x}'))\mathbf{t}\, J\, 0.25\, d\xi\, d\eta \quad (27)$$

where $\mathbf{x}'$ is obtained by projecting $\mathbf{x}$ to a *plane strain axis*, which is defined by its origin $(\mathbf{x}_{ps})$ and its direction $(\mathbf{v}_{ps})$. This integral now has a finite value and since the integral of the second term in parentheses is zero for a closed contour, there is no change to the original integral.

The coordinates of $\mathbf{x}'$ are computed by (see Figure 12):

$$\mathbf{x}' = \mathbf{x}_{ps} + \left[(\mathbf{x} - \mathbf{x}_{ps}) \cdot \mathbf{v}_{ps}\right] \mathbf{v}_{ps} \quad (28)$$

3 SYMMETRY

The implementation of symmetry follows the one introduced earlier for the 2D case. Depending on the number of specified symmetry planes there are one, three or seven mirrored patches (see Figure 13). The multiplication factors for the coordinates, tractions and displacements can be computed by:

$$\begin{aligned}
\mathbf{x}^n &= \mathbf{T}^n \mathbf{x} \\
\mathbf{t}^n &= \mathbf{T}^n \mathbf{t} \\
\mathbf{u}^n &= \mathbf{T}^n \mathbf{u}
\end{aligned} \quad (29)$$

where $\mathbf{T}^n$ are transformation matrices. For $n = 1$ we have for example

$$\mathbf{T}^1 = \begin{pmatrix} -1 & 0 & 0 \\ 0 & 1 & 0 \\ 0 & 0 & 1 \end{pmatrix} \quad (30)$$

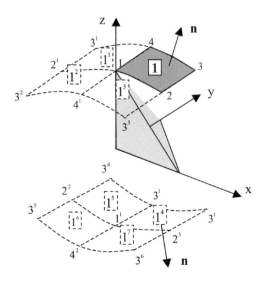

Figure 13 Explanation of symmetry.

As explained previously mirrored patches inherit the incidences of the original patches. For mirrored patches 1, 3, 4, and 6 the outward normal has to be reversed. A listing of function Symm, that determines the factors is shown[5]:

```
function [fac,rev]=Symm(nsym)
%-------------------------------------------
% generates information for symmetry
%
% Input:
% nsym  ... symmetry counter
%
% Output:
% fac   ... multiplication factors for displacements and coordinates
% rev   ... indicator for the reversal of the outward normal (0=no,1=yes)
%-------------------------------------------
rev=0;
if(nsym == 1) fac= [1,1,1];
elseif(nsym == 2) fac= [-1,1,1]; rev=1;
elseif(nsym == 3) fac= [-1,-1,1];
elseif(nsym == 4) fac= [1,-1,1]; rev=1;
elseif(nsym == 5) fac= [1,1,-1]; rev=1;
elseif(nsym == 6) fac= [-1,1,-1];
elseif(nsym == 7) fac= [1,-1,-1]; rev=1;
elseif(nsym == 8) fac= [-1,-1,-1]; endif
endfunction;
```

---

[5]For efficiency only the diagonals of the transformation matrix are stored.

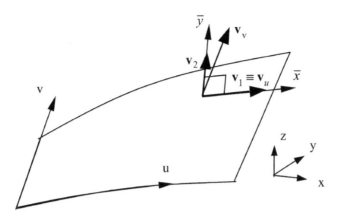

Figure 14 Determination of local axes for stress recovery.

4 MULTIPLE PATCHES

For multiple patches we usually expect that displacements are continuous where patches connect, whereas tractions can be discontinuous. We recall that the anchor points associated with the unknown parameters, computed using Greville abscisae, are also the collocation points. To ensure compatibility and a unique location of the collocation points we must ensure that the anchors are at the same location, when computed locally for each connecting NURBS patch, and have a unique number. The implementation follows the one for the 2D case, i.e. appropriate coefficients are added.

However, whereas in the FEM compatibility of displacements is essential for convergence this is not the case with the BEM. This means that we can place the collocation points (and the places where the unknowns are computed) slightly inside the NURBS patch. In this case the location of collocation points does not need to match, giving greater freedom in the simulation especially when dealing with trimmed surfaces. The method known as *discontinuous collocation* results in small discontinuities at the interface but convergence is not affected. More details can be found in [4].

5 POSTPROCESSING

As with plane problems we have two types of post processing. One for the results on the boundary, using stress recovery, and one for internal results.

5.1 Stress recovery

To obtain the stresses tangential and normal to the boundary we must establish an orthogonal local coordinate system.

First vectors tangential to the boundary as shown in Figure 14 are computed.

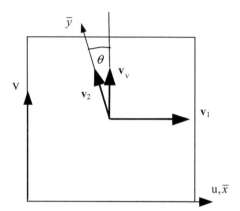

Figure 15 Relationship between $\bar{x}, \bar{y}$ and u, v coordinates.

The vectors are given by

$$\mathbf{v}_u = \frac{\partial \mathbf{x}}{\partial u} \cdot \frac{1}{J_u} \tag{31}$$

$$\mathbf{v}_v = \frac{\partial \mathbf{x}}{\partial v} \cdot \frac{1}{J_v}$$

where J_u, J_v are the stretch factors given by

$$J_u = \sqrt{\left(\frac{\partial x}{\partial u}\right)^2 + \left(\frac{\partial y}{\partial u}\right)^2 + \left(\frac{\partial z}{\partial u}\right)^2} \tag{32}$$

$$J_v = \sqrt{\left(\frac{\partial x}{\partial v}\right)^2 + \left(\frac{\partial y}{\partial v}\right)^2 + \left(\frac{\partial z}{\partial v}\right)^2}$$

Since vectors $\mathbf{v}_u$ and $\mathbf{v}_v$ are not orthogonal vectors, we establish an orthogonal system $\mathbf{v}_1$ and $\mathbf{v}_2$ where $\mathbf{v}_1 = \mathbf{v}_u$ and $\mathbf{v}_2$ is computed by a vector x-product.

Referring to Figure 15 the relationship between the orthogonal coordinates $\bar{x}$, $\bar{y}$ and local coordinates u, v is given by

$$\bar{x} = J_u \cdot u + J_v \cdot v \cdot \cos\theta \tag{33}$$

$$\bar{y} = J_v \cdot v \cdot \sin\theta$$

where

$$\cos\theta = \mathbf{v}_u \cdot \mathbf{v}_v, \quad \sin\theta = \mathbf{v}_v \cdot \mathbf{v}_2 \tag{34}$$

The inverse relationship between $\bar{x}, \bar{y}$ and local coordinates u,v is given by

$$u = \frac{1}{J_u}(\bar{x} - \cot\theta \cdot \bar{y}), \quad v = \frac{1}{J_v \sin\theta}\bar{y} \tag{35}$$

and the derivatives are

$$\frac{\partial u}{\partial \bar{x}} = \frac{1}{J_u}, \quad \frac{\partial u}{\partial \bar{y}} = \frac{1}{J_u} \cdot \cot \theta, \quad \frac{\partial v}{\partial \bar{y}} = \frac{1}{J_v \sin \theta} \tag{36}$$

The strains in the $\bar{x}, \bar{y}$ coordinate system are given by

$$\epsilon_{\bar{x}} = \frac{\partial u_{\bar{x}}}{\partial \bar{x}} = \left(\frac{\partial \mathbf{u}}{\partial u} \cdot \mathbf{v}_1 \right) \frac{\partial u}{\partial \bar{x}}$$

$$\epsilon_{\bar{y}} = \frac{\partial u_{\bar{y}}}{\partial \bar{y}} = \left(\frac{\partial \mathbf{u}}{\partial u} \cdot \mathbf{v}_2 \right) \frac{\partial u}{\partial \bar{y}} + \left(\frac{\partial \mathbf{u}}{\partial v} \cdot \mathbf{v}_2 \right) \frac{\partial v}{\partial \bar{y}} \tag{37}$$

$$\gamma_{\bar{y}} = \frac{\partial u_{\bar{x}}}{\partial \bar{y}} + \frac{\partial u_{\bar{y}}}{\partial \bar{x}} = \left(\frac{\partial \mathbf{u}}{\partial u} \cdot \mathbf{v}_1 \right) \frac{\partial u}{\partial \bar{y}} + \left(\frac{\partial \mathbf{u}}{\partial v} \cdot \mathbf{v}_1 \right) \frac{\partial v}{\partial \bar{y}} + \left(\frac{\partial \mathbf{u}}{\partial u} \cdot \mathbf{v}_2 \right) \frac{\partial u}{\partial \bar{x}}$$

Finally applying Hooke's law we have:

$$\sigma_{\bar{x}} = C_1 (\epsilon_{\bar{x}} + v\epsilon_{\bar{y}}) + C_2 t_{\bar{z}}, \quad \sigma_{\bar{y}} = C_1 (\epsilon_{\bar{y}} + v\epsilon_{\bar{x}}) + C_2 t_{\bar{z}}$$

$$\sigma_{\bar{z}} = t_{\bar{z}}, \quad \tau_{\bar{x}\bar{y}} = G\gamma_{\bar{x}\bar{y}} \tag{38}$$

$$\tau_{\bar{x}\bar{z}} = t_{\bar{x}}, \quad \tau_{\bar{y}\bar{z}} = t_{\bar{y}}$$

where G is the shear modulus and $t_{\bar{x}}$, $t_{\bar{y}}$ and $t_{\bar{z}}$ are the components of traction in the $\bar{x}, \bar{y}$ and $\bar{z}$ (normal to the surface) directions and

$$C_1 = \frac{E}{1 - v^2}, \quad C_1 = \frac{v}{1 - v} \tag{39}$$

5.2   Internal stress computation

For elasticity problems the stresses are collected in a pseudo-vector using the Voight notation and are given by:

$$\boldsymbol{\sigma}(\mathbf{y}) = \sum_{e=1}^{E} \Delta \mathbf{S}^e - \sum_{e=1}^{E} \Delta \mathbf{R}^e \tag{40}$$

where E is the number of patches and

$$\Delta \mathbf{S}^e = \int_0^1 \int_0^1 \mathbf{S} \cdot \mathbf{t} J \, du \, dv \tag{41}$$

$$\Delta \mathbf{R}^e = \int_0^1 \int_0^1 \mathbf{R} \cdot \mathbf{u} J \, du \, dv \tag{42}$$

The fundamental solutions $\mathbf{S}$ and $\mathbf{R}$ are given in the Appendix.

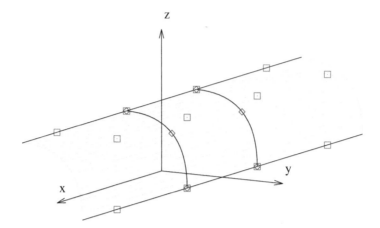

Figure 16 Discretization of an infinite tunnel into one finite patch and two infinite patches showing control points (red) and collocation points (blue).

6 TEST EXAMPLES

6.1 Infinite tunnel

The first example is to test the infinite plane strain elements. Figure 16 shows a discretization which exactly describes a quarter of the geometry of a circular tunnel with one finite patch and two infinite plane strain patches. Two symmetry planes have been assumed for the simulation. The tunnel is excavated in an infinite prestressed domain (virgin stress σ_0 in vertical direction 1, all other components zero). This is a pure Robin problem and the tractions are given as:

$$\mathbf{t} = \mathbf{n} \cdot \sigma_0 \tag{43}$$

where $\mathbf{n}$ is the outward normal.

For the purpose of the test Young's modulus was assumed to be 1 and Poisson's ratio 0. Also shown in the Figure are the collocation points for an isogeometric analysis (i.e. the same basis functions are used for the description of the geometry and the unknown). Since the unknown displacements are constant in the infinite direction a linear variation in this direction is sufficient. In the circumferential direction a NURBS of order 2 happens to be able to exactly describe the variation of displacements.

The maximum z-displacement computed is 2.005 compared with the theoretical value of 2.0. The maximum tangential stress, computed by stress recovery is 3.002 as compared with the exact value of 3.0. As in this case the geometry, the variation of the known values and the variation of the unknown is exactly represented, the error is only due to the integration error (the integration precision in this case was set to 10^{-3}).

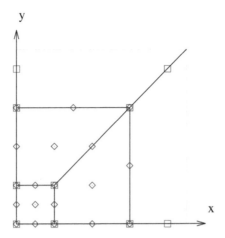

Figure 17   Discretization of half-space into three finite and 2 infinite (decay) patches showing control
points (red) and collocation points (blue).

6.2   Loading on infinite half-space

The second example is designed to test the decay infinite NURBS.

The discretization is shown in Figure 17 and consists of 3 finite NURBS patches
with $p=1$, $q=1$ for the geometry description and $p_d=2$, $q_d=2$ for the description of the displacements and 2 infinite *decay* NURBS patches with $q=1$ and $q_d=2$.
Symmetry conditions are applied as before so only 1/4 of the problem is discretized.
The problem has 57 degrees of freedom.

The NURBS patch at the center is loaded with a vertical distributed load of
100 kPa. The material properties are $E=10\,000$ kPa and $\nu=0$. The same problem
was solved with the isoparametric BEM in [6]. The theoretical maximum displacement (see [5]) is 22.444 mm. The result of the isogeometric analysis is 22.56 and the
one reported in reference [6] with a similar isoparametric mesh is 22.52[6].

7   EXAMPLES

7.1   Infinite tunnel in infinite domain near tunnel face

The first example relates to the simulation of a tunnel excavated in an infinite elastic
domain. It is the same as in the previous section except that the tunnel only extends
to infinity in one direction and is truncated by a tunnel face. The discretization into
4 finite (2 of them degenerate) and 2 infinite NURBS patches is shown in Figure 18.
Only one plane of symmetry has been specified.

---

[6] As pointed out in the quoted reference the result is slightly dependent on the location of the
nodes on the edge of the infinite patches pointing to infinity.

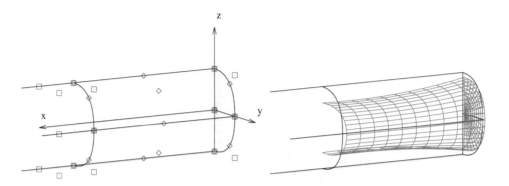

Figure 18 Discretization of tunnel into 4 finite and 2 infinite NURBS patches and displaced shape. Control points (red) and collocation points (blue) are shown for the first stage of refinement.

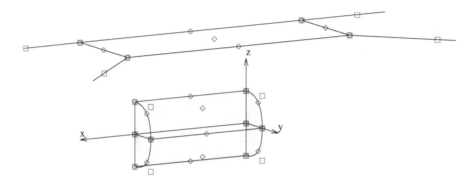

Figure 19 Definition of the tunnel in a semi-infinite domain. Collocation points (in blue) are shown for the first refinement.

A refinement of the geometry description (originally of order 1) to order 2 was made in the direction along the tunnel. Figure 18 shows the displaced shape of the tunnel. It can be seen that plane strain conditions are encountered about 1.5 diameters away from the tunnel face.

The maximum vertical displacement is 1.96, which is close to the theoretical value for the infinite tunnel. Further refinements did not change the result.

7.2 Finite tunnel in a semi-infinite domain

Next we analyze a tunnel of finite length in a semi-infinite domain described by decay infinite NURBS patches. The discretization into 7 finite NURBS patches and 3 infinite (decay) patches is shown in Figure 19.

One plane of symmetry was considered and the properties and the virgin stress field are the same as for the previous example. Figure 20 shows the displaced shape after the first refinement.

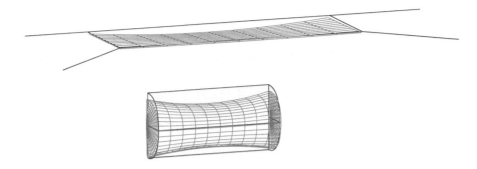

Figure 20 Displaced shape after the first refinement.

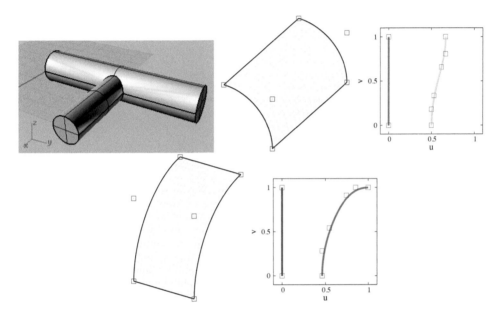

Figure 21 CAD model of tunnel branch and extracted NURBS surface and trimming information.

7.3 Branched tunnel

This example shows how geometrical data can be used directly from the CAD program without the need for mesh generation. Figure 21 shows the CAD model of a tunnel intersection and the geometrical information extracted from a file generated by the program Rhino. This information is used directly for the simulation as follows: The trimmed surfaces are used to model 1/4 of the problem as shown in Figure 22.

Two planes of symmetry (about the x-y and x-z planes) are assumed and infinite plane strain patches are used to simulate infinitely long tunnels. The tunnels are assumed to be excavated in an elastic pre-stressed ground (for more details see [3], [2] and [1]).

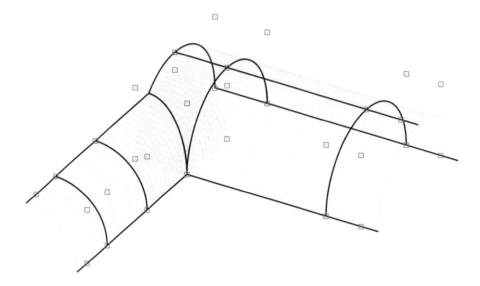

Figure 22 Discretisation of tunnel intersection problem into 5 finite and 2 infinite plane strain NURBS patches showing control points.

We analyze the tunnel intersection with the following properties.

- Elastic domain with $E = 1000$ MPa, $v = 0$
- Virgin stress: $\sigma_z = 1$ MPa compression, all other components zero
- Symmetry about x-z and x-y planes
- Single stage excavation

The locations of collocation points are computed in the local patch coordinate system using the Greville formula and then mapped into the global system.

It should be noted here that collocation points will only be in the same position if the basis functions and the parameters spaces of the trimming curves match along the intersection between two patches. Unfortunately, one can not rely on the fact that the trimming information supplied by the CAD program is such that parameter spaces of the curves match at the intersection.

We have learned earlier that parameter spaces are influenced by the entries in the Knot vector, so some manipulation of these may result in matching parameter spaces. If the parameter spaces can not be made to match within a certain tolerance then the remedy would be either to recalculate the parameters of the trimming curves (using points computed on the curves) or to use discontinuous collocation as mentioned. Fortunately in this case the location of the collocation points matched and continuous collocation could be used.

Another issue that needs to be addressed is the integration. Since NURBS patches are much bigger than Finite Elements a subdivision is required.

In the program, subdivision lines are generated automatically through collocation points. This is required for the weakly singular integration to work. To account for basis functions, which are not continuous over the patch subregions are also introduced

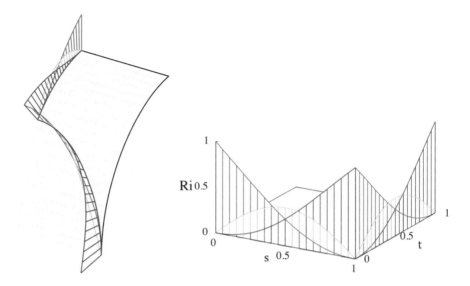

Figure 23 Basis functions before refinement in s,t and global coordinate system.

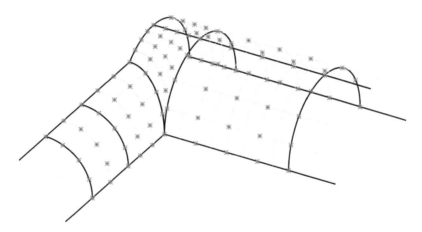

Figure 24 Location of collocation points and subdivision into integration regions for second refinement
stage.

at knots. Further subdivisions are automatically made by the program for the case
where the source point is close to the integration region, using the Quadtree method.

We follow the method of geometry independent approximation, i.e. we keep the
geometry description unchanged and only change the approximation of the unknown.
Figure 23 shows traces the basis functions defined in the s, t coordinate system and in
the x, y, z coordinate system before refinement for one of the trimmed surfaces.

For the analysis the order of the basis functions was elevated until convergence was
achieved. Figure 24 shows the location of the collocation points for orders $p = q = 4$

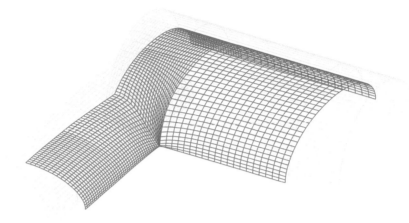

Figure 25 Deformed shape.

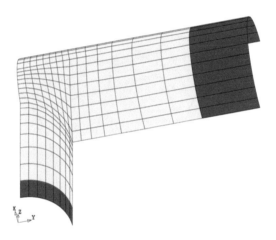

Figure 26 Mesh used for the conventional BEM analysis with isoparametric elements.

(291 degrees of freedom) and Figure 25 one result of the analysis namely the deformed shape.

To check the accuracy, the results are compared with a conventional BEM analysis using Serendipity functions for describing the geometry and the variation of the unknowns. Figure 26 shows the mesh used for the analysis with the simulation program BEFE. Two analyses were performed, one with linear and one with quadratic shape functions. The latter had 2895 unknowns.

The z-displacement along the trimming curve is shown in Figure 27 for the conventional BEM and the new approach.

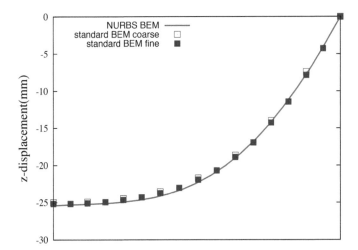

Figure 27 Variation of the vertical displacement along the trimming line, comparison of new method with isoparametric BEM.

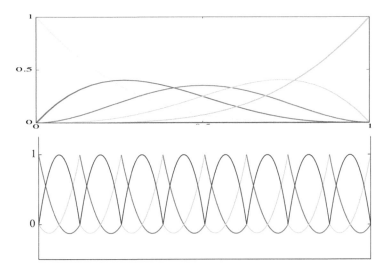

Figure 28 Comparison of the basis functions along the intersection of the two surfaces. Top: for NURBS based BEM, bottom: conventional BEM.

It can be seen that the conventional BEM results converge towards the results for the NURBS based BEM. The conventional BEM analysis used about 10 times more unknowns to achieve the same quality of results. The reason for this is that many boundary elements are required just to approximate the geometry.

If we compare the variation of the basis functions along the intersection line in Figure 28 then we can see that the NURBS basis functions are continuous whereas the Serendipity functions are only piecewise continuous.

8   CONCLUSIONS

Here the implementation of the 3-D NURBS based BEM was discussed. The main issue was the treatment of nearly singular and singular integrands. Care had to be taken to ensure the accuracy of integration, which is much more important than for the FEM, where a reduced integration can be even beneficial. In the BEM consequences can be quite severe if the quality of the integration is not maintained. The Gauss Quadrature applied is an old and trusted method, initially conceived for integrating polynomial functions and it has worked well for the FEM. However, since the integrands that we are dealing with are not polynomials and contain some nasty singular properties it is worth considering if there are better alternatives. Although some ideas have been published, they are not sufficiently convincing to abandon Gauss. The most compute intensive part of the BEM program is the numerical integration and since the evaluation of the NURBS basis functions and derivatives uses up more floating point operations as compared with Serendipity functions this is an important aspect to be considered for efficiency.

Although, as we have seen, it is much more complicated to program the BEM, there are considerable benefits to be gained for the user: Simpler definition of the geometry and much more accurate results with fewer degrees of freedom. As has been shown in the examples, using NURBS for the description of the geometry and for the approximation of the unknown result in a significant increase in accuracy and fewer unknowns. As pointed out, BEM is the only method that can lead to an achievement of a seamless integration of CAD and simulation.

So far we have only dealt with homogeneous and elastic domains and this would severely limit the application, especially in geomechanics. The reason for this restriction is that we have not considered body forces in the formulation. Body forces are introduced in the next stage. It will be seen that this results in some volume integrals and on first glance the advantage of the BEM, namely surface only discretization, seems to be lost. However, there is no reason to despair. The volume integrals are restricted to zones where nonlinear behavior occurs or where there are inclusions, there is no increase in the number of unknowns and as will be shown there is no need to provide a volume mesh. Thus the suitability of the BEM as a ideal companion to CAD is not affected.

BIBLIOGRAPHY

[1]   G. Beer. Mapped infinite patches for the NURBS based boundary element analysis in geomechanics. *Computers and Geotechnics*, 66:66–74, 2015.

[2]   G. Beer, B. Marussig, and J. Zechner. A simple approach to the numerical simulation with trimmed CAD surfaces. *Computer Methods in Applied Mechanics and Engineering*, 285:776–790, 2015.

[3]   Gernot Beer, Benjamin Marussig, Juergen Zechner, Christian Duenser, and Thomas-Peter Fries. Boundary Element Analysis with trimmed NURBS and a generalized IGA approach. In E. Oñate, J. Oliver, and A. Huerta, editors, *11th World Congress on Computational Mechanics (WCCM XI)*, 2014.

[4]   Benjamin Marussig, Jürgen Zechner, Gernot Beer, and Thomas-Peter Fries. Fast isogeometric boundary element method based on independent field approximation. *Computer*

Methods in Applied Mechanics and Engineering, 284(0):458–488, 2015. Isogeometric Analysis Special Issue.

[5]   S.P. Timoshenko and J.N. Goodier. *Theory of Elasticity*. McGraw-Hill, New York, NY, USA, 1970.

[6]   Ch. Duenser W. Moser and G. Beer. Mapped infinite elements for three-dimensional multi-region boundary element analysis. *International Journal for Numerical Methods in Engineering*, 61:317–328, 2004.

[7]   J.O. Watson. *Developments in Boundary Element Methods – 1*, volume 1, chapter Advanced implementation of the boundary element method for two- and threedimensional elastostatics. Applied Science, 1979.

Chapter 10

Stage 9: The boundary element method with volume effects

Gravitation is not responsible for people falling in love

J. Keppler

where we extend the capabilities of the BEM to simulate heterogeneous and non-linear problems.

1 INTRODUCTION

So far we we have assumed that loading only occurs on the boundary, i.e. we have neglected the influence of effects inside the domain. In addition, we have assumed in the computation of the fundamental solutions, that there is a linear relationship between stress and strain or flow and potential and that the domain is homogeneous, i.e. has the same properties everywhere.

This severely restricts the applicability of the method to practical problems. In applications in geomechanics for example the rock/soil mass will exhibit non-linear behavior during loading. In addition, the ground will not be homogeneous and may have inclusions with different material properties. Finally there may be some effects inside the domain, which need to be considered, for example volume change due to swelling.

As will be shown, the first two problems can be considered in an iterative way. Readers may be familiar with the initial stress method used in the FEM which involves a number of linear solutions with a residual right hand side, which tends to zero as the iteration converges. In plasticity for example the computation of the right hand side involves forces inside the domain which relate to the stresses being redistributed.

To extend the capabilities of the method we re-introduce volume effects that have been neglected so far. Following a standard that seems to have developed in the literature, we use an overdot to indicate an increment. This is convenient as it avoids the use of Δ but must not be confused with the use of an overdot for the first time derivative.

2 EFFECT OF BODY FORCES AND INITIAL STRAIN

Here we discuss two types of volume effects: Forces that occur inside the domain and initial strains. The first occur when dealing with inelastic behavior and heterogeneous

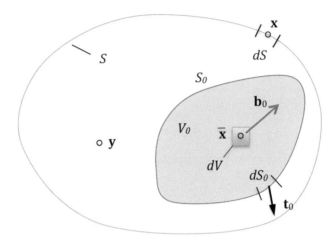

Figure 1 Explanation of the application of Betti's theorem with body forces.

domains, the latter when parts of the domain are subjected to a known volume increase (for example swelling).

2.1 Body forces

Here we investigate the effect of incremental body forces $\dot{\mathbf{b}}_0$, that occur inside a sub-domain V_0 as shown in Figure 1. The origin of the body forces may be due to in-elastic behavior or due to inclusions, as will be elaborated later.

Applying Betti's theorem, as explained in stage 6 we have to consider the **additional** work done by the displacements of load case 1 and the body forces of load case 2 on a small volume dV of the subdomain:

$$d\mathbf{W}_{12}^0 = \mathbf{U}(\mathbf{y}, \bar{\mathbf{x}}) \cdot \dot{\mathbf{b}}_0(\bar{\mathbf{x}}) \cdot dV_0 \tag{1}$$

If the body forces are related to initial stresses then we also have to consider the work done by the following tractions at the boundary of V_0:

$$\mathbf{t}_0 = \mathbf{n} \cdot \boldsymbol{\sigma}_0 \tag{2}$$

where $\mathbf{n}$ is the outward normal to S_0 and $\boldsymbol{\sigma}_0$ is the initial stress.

The additional work done is:

$$d\mathbf{W}_{12}^{0t} = \mathbf{U}(\mathbf{y}, \bar{\mathbf{x}}) \cdot \dot{\mathbf{t}}_0(\bar{\mathbf{x}}) \cdot dS_0 \tag{3}$$

The boundary integral equation with body forces can now be written as:

$$c \cdot \mathbf{u}(\mathbf{y}) = \int_S \mathbf{U}(\mathbf{y}, \mathbf{x}) \mathbf{t}(\mathbf{x}) dS + \int_{S_0} \mathbf{U}(\mathbf{y}, \bar{\mathbf{x}}) \dot{\mathbf{t}}_0(\bar{\mathbf{x}}) dS_0 \qquad (4)$$

$$- \int_S \mathbf{T}(\mathbf{y}, \mathbf{x}) \mathbf{u}(\mathbf{x}) dS + \int_{V_0} \mathbf{U}(\mathbf{y}, \bar{\mathbf{x}}) \dot{\mathbf{b}}_0(\bar{\mathbf{x}}) dV_0$$

We can see that an additional integral over volume V_0 occurs. For potential problems we have for a internal flow/volume $\dot{b}_0$:

$$c \cdot u(\mathbf{y}) = \int_S \mathbf{U}(\mathbf{y}, \mathbf{x}) t(\mathbf{x}) dS - \int_S \mathbf{T}(\mathbf{y}, \mathbf{x}) \, u(\mathbf{x}) dS + \int_{V_0} \mathbf{U}(\mathbf{y}, \bar{\mathbf{x}}) \dot{b}_0(\bar{\mathbf{x}}) dV \qquad (5)$$

The implementation of volume effects was first suggested by Telles [11] for the treatment of inelastic material behavior. He suggested to discretize the volume using cells, that are identical to finite elements. This means that the distribution of body forces inside the cell is approximated using basis functions. This results in an additional discretization effort but does not increase the number of degrees of freedom. Details of the implementation of a cell based inelastic solution can be found in [2]. A further development reported in [8] discusses the automatic generation of cells. The method was extended to the treatment of inclusions in [9].

The disadvantage of the cell method is that the approximation of the body force inside cells introduces errors into the solution, especially when the cell is transected by the elasto-plastic boundary.

Here we propose a different, more accurate way. Instead of using cells for discretizing the volume, we use a geometrical definition by NURBS and mapping. This means that no approximation of the body forces inside the plastic domain is required and most importantly, that the integration is carried out only in the domain. It will also be shown how the extent of the plastic domain can be automatically detected, so no additional work would be required by the user.

The ideas presented here are quite novel and some have not been verified. This is with the spirit of this book, that is meant as an impetus for readers to work in this exciting new area.

2.2 Effect of initial strain

Here we assume that known strains ϵ_0 are generated inside the domain. This would allow us to consider effects such as swelling.

The additional work done on a small volume dV is by the loads of load case 1 times the displacements of load case 2 is

$$d\mathbf{W}_{21}^0 = \mathbf{\Sigma}(\mathbf{y}, \bar{\mathbf{x}}) \cdot \epsilon_0(\bar{\mathbf{x}}) \cdot dV \qquad (6)$$

where $\mathbf{\Sigma}(\mathbf{y}, \bar{\mathbf{x}})$ is a matrix of fundamental solutions for stress, listed in the Appendix.

The integral equation is

$$c \cdot \mathbf{u}(\mathbf{y}) = \int_S \mathbf{U}(\mathbf{y}, \mathbf{x}) \mathbf{t}(\mathbf{x}) dS - \int_S \mathbf{T}(\mathbf{y}, \mathbf{x}) \mathbf{u}(\mathbf{x}) dS + \int_{V_0} \mathbf{\Sigma}(\mathbf{y}, \bar{\mathbf{x}}) \epsilon_0(\bar{\mathbf{x}}) dV \qquad (7)$$

2.3 Solution

The solution with volume effect involves an additional right hand side. The system of Equations to be solved is[1].

$$[\mathbf{T}]\{\mathbf{u}\} = \{\mathbf{F}\} + \{\mathbf{F}^0\} \tag{8}$$

where the subvectors of $\{\mathbf{F}^0\}$ due to initial strain are given by

$$\mathbf{F}_n^0 = \int_{V_0} \mathbf{\Sigma}(\mathbf{y}_n, \bar{\mathbf{x}}) \boldsymbol{\epsilon}_0(\bar{\mathbf{x}}) dV \tag{9}$$

For the inelastic solutions and for the consideration of inclusions the solution has to proceed iteratively with the first solution being:

$$[\mathbf{T}]\{\mathbf{u}\} = \{\mathbf{F}\} \tag{10}$$

For subsequent solution we compute incremental displacements $\{\dot{\mathbf{u}}\}$ from incremental body forces by

$$[\mathbf{T}]\{\dot{\mathbf{u}}\} = \{\dot{\mathbf{F}}^0\} \tag{11}$$

where

$$\dot{\mathbf{F}}_n^0 = \int_{V_0} \mathbf{U}(\mathbf{y}_n, \bar{\mathbf{x}}) \dot{\mathbf{b}}_0(\bar{\mathbf{x}}) dV(\bar{\mathbf{x}}) \tag{12}$$

3 IMPLEMENTATION FOR PLANE PROBLEMS

Here we discuss the practical implementation of volume effects for plane problems. We start with the definition of the subdomain V_0 and then proceed to the numerical evaluation of the volume integral.

3.1 Geometry definition

The first task is the description of the geometry of the subdomain V_0. There are various ways in which this can be done. One way would be to discretize the volume into cells. Here we propose a simpler, way, which is consistent with the isogeometric philosophy, i.e. that complex geometries can be defined with few parameters.

We propose to use the mapping method introduced earlier for trimmed surfaces. This means that the subdomain is defined by two NURBS curves and a linear interpolation between them.

We establish a local coordinate system s,t as shown in Figure 2 and perform all computations such as integration in this system and then map it to the global x, y system. Note that the local coordinate s ranging from 0 to 1 is the same as the local

---

[1] For simplicity we will restrict further discussions on pure Neumann or Robin BC. The extension to Dirichlet or mixed problems is straight forward.

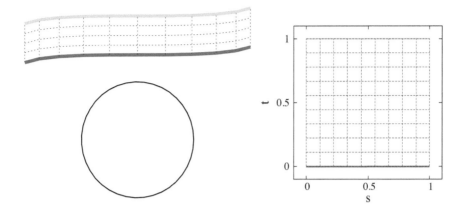

Figure 2   Definition of subdomain with 2 NURBS curves in left global and right local coordinate space.

coordinate u of the NURBS curve defining the boundary. The global coordinates of a point, $\mathbf{x}$, with the local coordinates s, t is given by:

$$\mathbf{x}(s,t) = (1-t)\cdot \mathbf{x}^{I}(s) + t\cdot \mathbf{x}^{II}(s) \tag{13}$$

where

$$\mathbf{x}^{I}(s) = \sum_{n=1}^{N^{I}} R_{n}^{I}(s)\cdot \mathbf{x}_{n}^{I}; \quad \mathbf{x}^{II}(s) = \sum_{n=1}^{N^{II}} R_{n}^{II}(s)\cdot \mathbf{x}_{n}^{II} \tag{14}$$

The superscript I relates to the bottom (red) curve and II to the top (green) curve and $\mathbf{x}_{n}^{I}$, $\mathbf{x}_{n}^{II}$ are control point coordinates. N^{I} and N^{II} is the number of control points and $R_{n}^{I}(s)$, $R_{n}^{II}(s)$ are NURBS basis functions.

The derivatives are given by:

$$\frac{\partial \mathbf{x}(s,t)}{\partial s} = (1-t)\cdot \frac{\partial \mathbf{x}^{I}(s)}{\partial s} + t\cdot \frac{\partial \mathbf{x}^{II}(s)}{\partial s}$$
$$\frac{\partial \mathbf{x}(s,t)}{\partial t} = -\mathbf{x}^{I}(s) + \cdot \mathbf{x}^{II}(s) \tag{15}$$

where

$$\frac{\partial \mathbf{x}^{I}(s)}{\partial s} = \sum_{n=1}^{N} \frac{\partial R_{n}^{I}(s)}{\partial s}\cdot \mathbf{x}_{n}^{I}$$
$$\frac{\partial \mathbf{x}^{II}(s)}{\partial s} = \sum_{n=1}^{N} \frac{\partial R_{n}^{II}(s)}{\partial s}\cdot \mathbf{x}_{n}^{II} \tag{16}$$

The Jacobi matrix of this mapping is:

$$\mathbf{J} = \begin{pmatrix} \dfrac{\partial x}{\partial s} & \dfrac{\partial y}{\partial s} \\ \dfrac{\partial x}{\partial t} & \dfrac{\partial y}{\partial t} \end{pmatrix} \tag{17}$$

and the Jacobian is $J = |\mathbf{J}|$

3.2 Computation of the volume integral

We use Gauss Quadrature to compute the volume integral. For this we introduce local coordinates ξ, η with the limits -1 to 1. The transformation of s, t to ξ, η coordinates is given by:

$$s = 0.5(\xi + 1); \quad t = 0.5(\eta + 1) \tag{18}$$

The Jacobian of this transformation is $J_s = 0.25$.
The integral for the body force can now be written as[2]:

$$\mathbf{F}_n^0 = \int_{-1}^{1} \int_{-1}^{1} \mathbf{U}(\mathbf{y}_n, \bar{\mathbf{x}}(\xi, \eta)) \dot{\mathbf{b}}_0(\bar{\mathbf{x}}(\xi, \eta)) J \cdot J_s \cdot d\xi \, d\eta \tag{19}$$

Applying Gauss integration we have:

$$\mathbf{F}_n^0 = \sum_{m=1}^{M} \sum_{n=1}^{N} \mathbf{U}(\mathbf{y}_n, \bar{\mathbf{x}}(\xi_m, \eta_n)) \dot{\mathbf{b}}_0(\bar{\mathbf{x}}(\xi_m, \eta_n)) J \cdot J_s \cdot W_m \cdot W_n \tag{20}$$

where M, N are the number of integration points in ξ and η directions. To determine the number of Gauss points necessary for an accurate integration we consider that whereas there is usually a moderate variation of body force, the Kernel $\mathbf{U}$ is $O(\ln r)$ so the number of integration points has to be increased if V_0 is close to the boundary.
In the case of initial strains we have

$$\mathbf{F}_n^0 = \sum_{m=1}^{M} \sum_{n=1}^{N} \boldsymbol{\Sigma}(\mathbf{y}_n, \bar{\mathbf{x}}(\xi_m, \eta_n)) \dot{\boldsymbol{\epsilon}}_0(\bar{\mathbf{x}}(\xi_m, \eta_n)) J \cdot J_s \cdot W_m \cdot W_n \tag{21}$$

Since the Kernel $\boldsymbol{\Sigma}$ is $O(\frac{1}{r})$ more Gauss points will be required when the subdomain is near the boundary.

4 IMPLEMENTATION FOR 3-D PROBLEMS

Here we extend the implementation to 3-D.

---

[2]It is assumed here that the dimension of the problem in the direction out of the plane is 1. For a plane stress problem the expression has to be multiplied with the thickness.

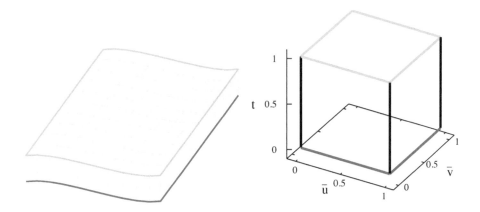

Figure 3 Definition of subdomain with 2 NURBS surfaces in left global and right local coordinate space.

4.1 Geometry definition

The volume is defined by 2 NURBS surfaces and a linear interpolation between them.

We establish a local coordinate system $\bar{u}, \bar{v}, t$ as shown in Figure 3 and perform all computations such as integration in this system and then map it to the global x, y, z system. The global coordinates of a point, $\mathbf{x}$, with the local coordinates $\bar{u}, \bar{v}, t$ is given by:

$$\mathbf{x}(\bar{u}, \bar{v}, t) = (1 - t) \cdot \mathbf{x}^I(\bar{u}, \bar{v}) + t \cdot \mathbf{x}^{II}(\bar{u}, \bar{v}) \tag{22}$$

where the coordinates on the bounding surfaces are given by:

$$\mathbf{x}^I(\bar{u}, \bar{v}) = \sum_{n=1}^{N^I} R_n^I(\bar{u}, \bar{v}) \cdot \mathbf{x}_n^I; \quad \mathbf{x}^{II}(\bar{u}, \bar{v}) = \sum_{n=1}^{N^{II}} R_n^{II}(\bar{u}, \bar{v}) \cdot \mathbf{x}_n^{II} \tag{23}$$

The superscript I relates to the bottom and II to the top surface and $\mathbf{x}_n^I$, $\mathbf{x}_n^{II}$ are control point coordinates. N^I and N^{II} is the number of control points and $R_n(\bar{u}, \bar{v})$ are tensor products of NURBS basis functions. The surfaces may be trimmed and in this case a mapping from u, v to $\bar{u}, \bar{v}$ system is required. For untrimmed surfaces $u = \bar{u}$ and $v = \bar{v}$.

The derivatives are:

$$\begin{aligned}
\frac{\partial \mathbf{x}(\bar{u}, \bar{v}, t)}{\partial \bar{u}} &= (1 - t) \cdot \frac{\partial \mathbf{x}^I(\bar{u}, \bar{v})}{\partial \bar{u}} + t \cdot \frac{\partial \mathbf{x}^{II}(\bar{u}, \bar{v})}{\partial \bar{u}} \\[2mm]
\frac{\partial \mathbf{x}(\bar{u}, \bar{v}, t)}{\partial \bar{v}} &= (1 - t) \cdot \frac{\partial \mathbf{x}^I(\bar{u}, \bar{v})}{\partial \bar{v}} + t \cdot \frac{\partial \mathbf{x}^{II}(\bar{u}, \bar{v})}{\partial \bar{v}} \\[2mm]
\frac{\partial \mathbf{x}(\bar{u}, \bar{v}, t)}{\partial t} &= -\mathbf{x}^I(\bar{u}, \bar{v}) + \mathbf{x}^{II}(\bar{u}, \bar{v})
\end{aligned} \tag{24}$$

The Jacobi matrix of this mapping is :

$$
\mathbf{J} = \begin{pmatrix} \dfrac{\partial x}{\partial \bar{u}} & \dfrac{\partial y}{\partial \bar{u}} & \dfrac{\partial z}{\partial \bar{u}} \\[2ex] \dfrac{\partial x}{\partial \bar{v}} & \dfrac{\partial y}{\partial \bar{v}} & \dfrac{\partial z}{\partial \bar{v}} \\[2ex] \dfrac{\partial x}{\partial t} & \dfrac{\partial y}{\partial t} & \dfrac{\partial z}{\partial t} \end{pmatrix}
\tag{25}
$$

and the Jacobian is $J = |\mathbf{J}|$.

4.2 Computation of the volume integral

We use Gauss Quadrature to compute the volume integral. For this we introduce local coordinates ξ, η, ζ with the limits -1 to 1. The transformation of u, v, t to ξ, η, ζ coordinates is given by:

$$
u = 0.5(\xi + 1); \quad v = 0.5(\eta + 1); \quad t = 0.5(\zeta + 1)
\tag{26}
$$

The Jacobian of this transformation is $J_s = 0.125$. The integral for the body force can now be written as:

$$
\mathbf{F}_n^0 = \int_{-1}^{1} \int_{-1}^{1} \int_{-1}^{1} \mathbf{U}(\mathbf{y}_n, \bar{\mathbf{x}}(\xi, \eta, \zeta)) \dot{\mathbf{b}}_0 \, (\bar{\mathbf{x}}(\xi, \eta, \zeta)) J \cdot J_s \cdot d\xi d\eta d\zeta
\tag{27}
$$

Applying Gauss integration we have:

$$
\mathbf{F}_n^0 = \sum_{m=1}^{M} \sum_{n=1}^{N} \sum_{l=1}^{L} \mathbf{U}(\mathbf{y}_n, \bar{\mathbf{x}}(\xi_m, \eta_n, \zeta_l)) \dot{\mathbf{b}}_0 \, (\bar{\mathbf{x}}(\xi_m, \eta_n, \zeta_l)) J \cdot J_s \cdot W_m \cdot W_n \cdot W_l
\tag{28}
$$

where M, N and L are the number of integration points in ξ, η and ζ directions. To determine the number of Gauss points necessary for an accurate integration we consider that whereas there usually is a moderate variation of body force, the Kernel $\mathbf{U}$ is $O(\frac{1}{r})$ so the number of integration points has to be increased if V_0 is close to the boundary.

For the initial strains we have

$$
\mathbf{F}_n^0 = \sum_{m=1}^{M} \sum_{n=1}^{N} \sum_{l=1}^{L} \mathbf{\Sigma}(\mathbf{y}_n, \bar{\mathbf{x}}(\xi_m, \eta_n, \zeta_l)) \dot{\boldsymbol{\epsilon}}_0 \, (\bar{\mathbf{x}}(\xi_m, \eta_n, \zeta_l)) J \cdot J_s \cdot W_m \cdot W_n \cdot W_l
\tag{29}
$$

Since the Kernel $\mathbf{\Sigma}$ is $O(\frac{1}{r^2})$ more Gauss points will be required when the subdomain is near the boundary.

5 ITERATIVE SOLUTION ALGORITHM

The iterative solution starts with a first solution $\{\mathbf{u}\}_0$, that neglects volume effects:

$$
[\mathbf{T}]\{\mathbf{u}\}_0 = \{\mathbf{F}\}
\tag{30}
$$

Next we compute a new right hand side with the body forces $\dot{\mathbf{b}}_0$ generated inside subdomain V_0. The sub-vector for the collocation point n and iteration i can be computed using Equation (20) and then assembled into the global vector $\{\mathbf{F}^0\}_i$.

With this new right hand side a correction to $\{\mathbf{u}\}$ is computed by

$$[\mathbf{T}]\{\dot{\mathbf{u}}\}_i = \{\mathbf{F}^0\}_i \tag{31}$$

The iteration continues until the norm of the residual is smaller than a given tolerance. The final solution is composed of the initial solution and the sum of all corrections:

$$\{\mathbf{u}\} = \{\mathbf{u}\}_0 + \{\dot{\mathbf{u}}\}_1 + \{\dot{\mathbf{u}}\}_2 + \cdots \tag{32}$$

We apply this algorithm to two problems: One where the domain is not homogeneous and one where there is not a linear relationship between stress and strain (inelastic behavior). In other words, we attempt to modify the results for an elastic, homogeneous domain, which were the assumptions for the fundamental solutions.

6   INCLUSIONS

Here we propose an iterative method to deal with inhomogeneous domains. However, we will restrict ourselves to the piecewise homogeneous case. This means that we assume that the inhomogeneities can be considered as inclusions, which have different elastic properties to the ones assumed for the fundamental solutions. Consider the example of a tunnel with a zone that has different material properties (E_2, ν_2) to the ones assumed for the fundamental solution (E_1, ν_1) (see Figure 2).

The proposed method proceeds as follows:

Obtain first a solution without the inclusion, i.e. assuming a Neumann problem solve:

$$[\mathbf{T}]\{\mathbf{u}\}_0 = \{\mathbf{F}\} \tag{33}$$

After the solution compute the strain at a point $\mathbf{y}$ inside the inclusion by:

$$\boldsymbol{\epsilon}(\mathbf{y}) = \sum_{e=1}^{E} \triangle\bar{\mathbf{S}}^e - \sum_{e=1}^{E} \triangle\bar{\mathbf{R}}^e \tag{34}$$

where E is the number of NURBS patches describing the boundary and for example for plane problems

$$\triangle\bar{\mathbf{S}}^e = \sum_{m=1}^{M} \bar{\mathbf{S}}(\mathbf{y}, \mathbf{x}(\xi_m(\mathbf{u}))\mathbf{t}(\mathbf{u}_m) \cdot J \cdot W_m \tag{35}$$

and

$$\Delta\bar{\mathbf{R}}^e = \sum_{m=1}^{M} \bar{\mathbf{R}}(\mathbf{y}, \mathbf{x}(\xi_m(\mathbf{u}))\mathbf{u}(\mathbf{u}_m) \cdot J \cdot W_m \tag{36}$$

where M is the number of Gauss points and J is the Jacobian of the mapping from local ξ coordinates to global coordinate. The derived fundamental solutions $\bar{\mathbf{R}}$ and $\bar{\mathbf{S}}$ are presented in the Appendix.

The stresses in the domain are:

$$\boldsymbol{\sigma} = \mathbf{C}_1 \cdot \boldsymbol{\epsilon} \tag{37}$$

where $\mathbf{C}_1$ is the elasticity matrix with the properties, that were used to compute the fundamental solution (E_1, ν_1). However, this is incorrect for the inclusion, where the stresses should have been computed using

$$\boldsymbol{\sigma} = \mathbf{C}_2 \cdot \boldsymbol{\epsilon} \tag{38}$$

where $\mathbf{C}_2$ is the elasticity matrix computed with the material properties of the inclusion (E_2, ν_2). A correction to the solution has now to be made. We compute a *residual stress* increment inside the inclusion:

$$\dot{\boldsymbol{\sigma}}^0 = (\mathbf{C}_1 - \mathbf{C}_2) \cdot \boldsymbol{\epsilon} \tag{39}$$

The stress has to be converted to a body force acting inside V_0 and a traction acting on its boundary S_0.

To compute the body force we recall the equilibrium equation:

$$b_j = -\frac{\partial \sigma_{jk}}{\partial x_k} \tag{40}$$

Changing to vector notation the residual body force vector, $\dot{\mathbf{b}}_0$, is computed by:

$$\dot{\mathbf{b}}_0 = - \begin{pmatrix} \dfrac{\partial \dot{\sigma}_x^0}{\partial x} + \dfrac{\partial \dot{\tau}_{xy}^0}{\partial y} + \dfrac{\partial \dot{\tau}_{xz}^0}{\partial z} \\[2ex] \dfrac{\partial \dot{\tau}_{xy}^0}{\partial x} + \dfrac{\partial \dot{\sigma}_y^0}{\partial y} + \dfrac{\partial \dot{\tau}_{yz}^0}{\partial z} \\[2ex] \dfrac{\partial \dot{\tau}_{xz}^0}{\partial x} + \dfrac{\partial \dot{\tau}_{yz}^0}{\partial y} + \dfrac{\partial \dot{\sigma}_z^0}{\partial z} \end{pmatrix} \tag{41}$$

The traction on the boundary S_0 is:

$$\mathbf{t}_0 = \mathbf{n} \cdot \dot{\boldsymbol{\sigma}}^0 \tag{42}$$

The following equation is solved for an increment in displacement:

$$[\mathbf{T}]\{\dot{\mathbf{u}}\}_1 = \{\mathbf{F}^0\}_1 + \{\mathbf{F}^{t0}\}_1 \tag{43}$$

where $\{\mathbf{F}^0\}_1$ is computed using Equation (20) or (28) and $\{\mathbf{F}^{t0}\}_1$ is computed by applying Gauss integration over the NURBS curves/surfaces defining the inclusion and the connecting lines/surfaces.

Using Equation (34), considering that the increment in tractions $\dot{\mathbf{t}}$ is zero and that the increment in body force $\dot{\mathbf{b}}_0$ is not zero we compute the strain increment:

$$\dot{\boldsymbol{\epsilon}}_1(\mathbf{y}) = -\sum_{e=1}^{E} \Delta\bar{\mathbf{R}}^e + \dot{\bar{\mathbf{S}}}_0 \tag{44}$$

where for plane problems $\dot{\bar{\mathbf{S}}}_0$ is computed by

$$\dot{\bar{\mathbf{S}}}_0 = \sum_{m=1}^{M}\sum_{l=1}^{L} \bar{\mathbf{S}}(\mathbf{y}, \bar{\mathbf{x}}(\xi_m, \eta_l)) \dot{\mathbf{b}}_0(\bar{\mathbf{x}}(\xi_m, \eta_l)) J \cdot J_s \cdot W_m \cdot W_l \tag{45}$$

With the computed strain increment we determine a new residual stress increment and body force. The iteration continues until the residual becomes smaller than a specified tolerance.

6.1   Computation of body force

For the computation of the body force the derivatives of the residual stresses are required. These are first computed in the local coordinate system and then expressed in the global coordinate system using the inverse of the Jacobi matrix as explained in stage 4.

For example, for plane problems the global derivatives of stress σ_x can be computed by:

$$\sigma_{x,x} = \mathbf{J}^{-1} \cdot \sigma_{x,s} \tag{46}$$

where

$$\sigma_{x,x} = \begin{pmatrix} \dfrac{\partial\sigma_x}{\partial x} \\ \dfrac{\partial\sigma_x}{\partial y} \end{pmatrix}; \quad \sigma_{x,s} = \begin{pmatrix} \dfrac{\partial\sigma_x}{\partial s} \\ \dfrac{\partial\sigma_x}{\partial t} \end{pmatrix} \tag{47}$$

The local derivatives are computed using finite differences. There are basically 2 ways to approach this: Creating a finite difference template for each Gauss point with a constant spacing or to use the Gauss points as finite difference points with non-uniform spacing. The former is obviously more expensive as for plane problems 4 and for 3-D problems 8 evaluations of stress are required for each Gauss point. The disadvantage of the second option is that Gauss points may be widely spaced, which could affect the accuracy.

Central finite difference template   For this we establish a finite difference template with spacing h as shown in Figure 4 for point i, j in the local coordinate space (s, t).

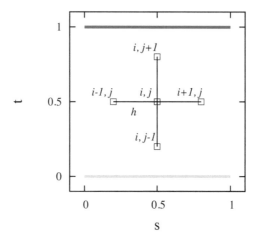

Figure 4 Central difference template for computing the derivatives.

We define 4 additional points, whose coordinates are:

$$
\begin{aligned}
s_{i+1,j} &= s_{i,j} + h; & t_{i+1,j} &= t_{i,j} \\
s_{i-1,j} &= s_{i,j} - h; & t_{i-1,j} &= t_{i,j} \\
s_{i,j+1} &= s_{i,j}; & t_{i,j+1} &= t_{i,j} + h \\
s_{i,j-1} &= s_{i,j}; & t_{i,j-1} &= t_{i,j} - h
\end{aligned}
\tag{48}
$$

where h is a small distance, chosen by experience. Care has to be taken that h is not chosen is such a way that the point falls outside the region of integration.

The derivatives of the stress Pseudo-vector at i,j are then given by:

$$
\frac{\partial \sigma^0}{\partial s} = \frac{\sigma^0_{i+1,j} - \sigma^0_{i-1,j}}{2h}
\tag{49}
$$

$$
\frac{\partial \sigma^0}{\partial t} = \frac{\sigma^0_{i,j+1} - \sigma^0_{i,j-1}}{2h}
$$

where for example $\sigma^0_{i+1,j}$ etc. are the stresses computed at the locations with the local coordinates $s_{i+1,j}$, $t_{i+1,j}$. The global coordinates coordinates of the points are obtained by mapping.

The method can be easily extended to 3-D problems. The central difference template has 8 additional points in this case.

Using finite differences with Gauss points Gauss points are spaced unequally so the finite difference formulae have to be modified. In addition we can no longer use a central finite difference for Gauss points that do not have other Gauss points on the side or above/below. For these points we must use a mix of central, forward or backward finite differences.

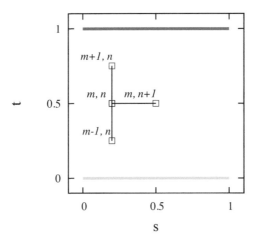

Figure 5 Computing the derivatives of σ^0 at a Gauss point.

As an example we show the evaluation of the derivatives for Gauss point (m, n) near the left boundary (see Figure 5):

$$\frac{\partial \sigma^0}{\partial s} = \frac{\sigma^0_{m,n+1} - \sigma^0_{m,n}}{d_s}$$

$$\frac{\partial \sigma^0}{\partial t} = \frac{\sigma^0_{m+1,n} - \sigma^0_{m-1,n}}{d_t}$$

(50)

where d_s is the distance between two Gauss points in s direction and d_t is the distance between the two adjacent Gauss points in t direction. The method can be easily extended to 3-D.

6.2 Steps in the analysis

The analysis considering inclusions proceeds with the following steps:

1 Solve the linear problem
2 Specify the geometry of the inclusion
3 Determine the number of Gauss points for the volume integration
4 For each Gauss point do
    Calculate the strain and the residual stress using (39)
    Calculate the local derivatives of the stress components using (49) or (50)
    Calculate the global derivatives with (46)
    Calculate the body force with (41)
5 Determine the new right hand side
6 Solve for increments in displacements
7 If the norm of the increments is below a specified tolerance finish otherwise repeat steps 4–6

7 INELASTIC BEHAVIOR

Inelastic behavior occurs, when the stresses reach the elastic limit, determined by a yield condition $F(\sigma)$. Inelastic behavior is often referred to as elasto-plastic behavior or plasticity stemming from its origin in the simulation of the behavior of metals. The term does not describe well the inelastic behavior of soils and rocks but is often used there. For simplicity we will refer to inelastic behavior as plasticity from now on.

We develop the integral equations for plasticity next but to keep the equations short we revert to tensor notation. We recall the relationship between stresses and strains:

$$\dot{\sigma}_{jk} = C_{jklm}\dot{\epsilon}^e_{lm} \tag{51}$$

where $\dot{\epsilon}^e_{lm}$ is the elastic strain increment and C_{jklm} is the elasticity tensor.

If the elastic limit has been reached (i.e. when $F(\sigma) = 0$), the total strain increment consists the sum of an elastic and a plastic part:

$$\dot{\epsilon}_{lm} = \frac{1}{2}\left(\frac{\partial \dot{u}_l}{\partial x_m} + \frac{\partial \dot{u}_m}{\partial x_l}\right) = \dot{e}^\epsilon_{lm} + \dot{\epsilon}^p_{lm} \tag{52}$$

where $\dot{\epsilon}^p_{lm}$ is the plastic strain increment and $\dot{u}_l$ denotes displacement increments.

Substitution into Equation (51) gives:

$$\dot{\sigma}_{jk} = C_{jklm}(\dot{\epsilon}_{lm} - \dot{\epsilon}^p_{lm}) \tag{53}$$

or

$$\dot{\sigma}_{jk} = \dot{\sigma}^e_{jk} - \dot{\sigma}^p_{jk} \tag{54}$$

where

$$\dot{\sigma}^e_{jk} = C_{jklm}\dot{\epsilon}_{lm} \tag{55}$$

and

$$\dot{\sigma}^p_{jk} = C_{jklm}\dot{\epsilon}^p_{lm} \tag{56}$$

Recall the differential equation of elasticity:

$$\frac{\partial \dot{\sigma}_{jk}}{\partial x_k} + b_j = 0 \tag{57}$$

Substitution of Equation (54) gives

$$\frac{\partial \dot{\sigma}^e_{jk}}{\partial x_k} - \frac{\partial \dot{\sigma}^p_{jk}}{\partial x_k} + b_j = 0 \tag{58}$$

or

$$\frac{\partial \dot{\sigma}^e_{jk}}{\partial x_k} + \dot{b}^p_j + b_j = 0 \tag{59}$$

where the plastic body force $\dot{b}^p_j = -\frac{\partial \dot{\sigma}^p_{jk}}{\partial x_k}$ has been introduced.

If the plastic zone V_0 extends to the boundary, then additional tractions have to be considered. The stresses on the boundary are related to the tractions by:

$$n_k \dot{\sigma}_{jk} = \dot{t}_j \tag{60}$$

or

$$n_k(\dot{\sigma}^e_{jk} - \dot{\sigma}^p_{jk}) = \dot{t}_j \tag{61}$$

In terms of $\dot{\sigma}^e_{jk}$ we have

$$n_k \dot{\sigma}^e_{jk} = \dot{t}_j + \dot{t}^p_j \tag{62}$$

where

$$\dot{t}^p_j = n_k \dot{\sigma}^p_{jk} \tag{63}$$

The integral equation can now be written in matrix notation as:

$$\mathbf{c}(\mathbf{y})\dot{\mathbf{u}}(\mathbf{y}) + \int_S \mathbf{T}(\mathbf{y},\mathbf{x})\dot{\mathbf{u}}(\mathbf{x})dS(\mathbf{x}) = \int_S \mathbf{U}(\mathbf{y},\mathbf{x})(\dot{\mathbf{t}}(\mathbf{x}) + \dot{\mathbf{t}}_p(\mathbf{x}))dS(\mathbf{x}) \tag{64}$$
$$+ \int_V \mathbf{U}(\mathbf{y},\bar{\mathbf{x}})\dot{\mathbf{b}}_p(\bar{\mathbf{x}})dV(\bar{\mathbf{x}})$$

where changing to vector notation we have:

$$\dot{\mathbf{t}}_p = \mathbf{n} \cdot \dot{\boldsymbol{\sigma}}^p \tag{65}$$

7.1 Yield conditions

The *yield condition* or *yield function* determines the limit of elastic straining. It is a function of stress and material constants and is also referred to as *yield surface*, since it can be visualized as a surface in principal stress space. The elastic limit is reached when $F(\boldsymbol{\sigma}) = 0$. For $F(\boldsymbol{\sigma}) < 0$ the state is elastic. A stress state that obeys $F(\boldsymbol{\sigma}) = 0$ is sometimes referred to as being *on the yield surface*.

A great number of yield conditions have been proposed for different materials (see for example [3] and [7]). Here we only discuss two of them: von Mises criterion for metals and the Mohr-Coulomb condition for soils. In addition we will restrict ourselves here to perfect plasticity[3].

The *von Mises* criterion is given by

$$F(\boldsymbol{\sigma}) = \sigma_{eq} - \sigma_Y \tag{66}$$

where in terms of the principal stresses $\sigma_1, \sigma_2, \sigma_3$:

$$\sigma_{eq} = \sqrt{0.5\left[(\sigma_1 - \sigma_2)^2 + (\sigma_2 - \sigma_3)^2 + (\sigma_3\sigma_1)^2\right]} \tag{67}$$

and σ_Y is the yield stress.

The *Mohr-Coulomb* yield condition is given by

$$F(\boldsymbol{\sigma}) = \frac{\sigma_1 + \sigma_3}{2}\sin\phi - \frac{\sigma_1 - \sigma_3}{2} - c \cdot \cos\phi \tag{68}$$

where c is the cohesion and ϕ is the angle of friction.

7.2 Determination of plastic strain increment

The next step is to compute the plastic strain increment. For this we introduce a *plastic potential* $Q(\boldsymbol{\sigma})$ that is used to determine the direction of plastic straining (*flow law*). The plastic potential is similar to the yield condition. If $Q(\boldsymbol{\sigma}) \equiv F(\boldsymbol{\sigma})$ then this is known as an *associative flow law*[4].

There are two approaches for determining the plastic strain increment. Elastoplasticity and viscoplasticity. In the first the stresses are not allowed to violate the Yield function, i.e. $F(\boldsymbol{\sigma}) \leqslant 0$ and this implies that stresses may need to be corrected to lie on the yield surface after an increment. In visco-plasticity we assume that plastic straining is a time dependent problem, i.e. $F(\boldsymbol{\sigma}) > 0$ is allowed but it is assumed that the stresses will obey $F(\boldsymbol{\sigma}) = 0$ after time has passed.

Visco-plasticity   In visco-plasticity we specify a visco-plastic strain rate as

$$\frac{\partial \boldsymbol{\epsilon}^{vp}}{\partial t} = \frac{1}{\eta}\Phi(F)\frac{\partial Q}{\partial \boldsymbol{\sigma}} \tag{69}$$

where η is a viscosity parameter and

$$\Phi(F) = 0 \quad \text{for } F < 0 \tag{70}$$

$$\Phi(F) = F \quad \text{for } F > 0 \tag{71}$$

---

[3]This means that the material parameters are constant, i.e. hardening and softening is not considered.
[4]In soil mechanics the flow law can be visualized as controlling the dilatancy, i.e. the volume increase during plastic straining.

The visco-plastic strain increment during a time increment Δt can be computed by an explicit scheme:

$$\dot{\epsilon}^{vp} = \frac{\partial \epsilon^{vp}}{\partial t} \cdot \Delta t \tag{72}$$

Δt can not be chosen freely and if chosen too large, oscillatory behavior will occur in the solution. Suitable time step values can be found in [1]. The visco-plastic stress increment is given by:

$$\dot{\sigma}^{vp} = \mathbf{C} \cdot \dot{\epsilon}^{vp} \tag{73}$$

Elasto-plasticity The increment in stress $\dot{\sigma}$, obtained after an iteration step can be written as:

$$\dot{\sigma} = \mathbf{C} \cdot \dot{\epsilon}^e = \mathbf{C} \cdot (\dot{\epsilon} - \dot{\epsilon}^p) \tag{74}$$

where $\dot{\epsilon}$ is the total strain increment. The plastic strain increment $\dot{\epsilon}^p$ is defined by:

$$\dot{\epsilon}^p = \lambda \frac{\partial Q}{\partial \sigma} \tag{75}$$

and after substitution of (75) into (74):

$$\dot{\sigma} = \mathbf{C} \cdot (\dot{\epsilon} - \lambda \frac{\partial Q}{\partial \sigma}) \tag{76}$$

The plastic multiplier λ is computed from the condition that $F(\sigma) \leqslant 0$ always. If $F = 0$ at the start of the increment then we have to ensure that $F \leqslant 0$ and this means that any positive change in F during loading has to be zero:

$$dF = \frac{\partial F}{\partial \sigma} \cdot \dot{\sigma} = 0 \tag{77}$$

Substituting (76) into (77) and solving for λ we obtain:

$$\lambda = \frac{1}{\beta} \frac{\partial F}{\partial \sigma} \mathbf{C} \cdot \dot{\epsilon} \tag{78}$$

where

$$\beta = \left(\frac{\partial F}{\partial \sigma} \right)^T \mathbf{C} \cdot \frac{\partial Q}{\partial \sigma} \tag{79}$$

In a numerical solution of equation (77) we assume a constant value of $\frac{\partial F}{\partial \sigma}$ for the increment and this means that the value of λ computed by (78) will be only an approximation. Therefore the computed stress state may not be on the yield surface and a correction has to be made. The discussion of such *return algorithms* is beyond

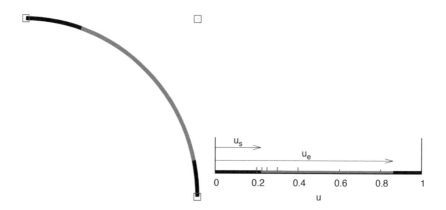

Figure 6 Explanation on how to determine the extent of the plastic zone on the boundary: NURBS patch with plastic zone marked in red (left in global coordinate space and right in local coordinate space). Also indicated in pink is the method used to detect the start of the plastic zone.

the scope of the book and the interested reader is referred to [5] and [10]. The plastic stress increment is given by:

$$\dot{\sigma}^p = \mathbf{C} \cdot \dot{\epsilon}^p \tag{80}$$

8 IMPLEMENTATION OF PLASTICITY FOR PLANE PROBLEMS

Here we discuss the implementation of inelastic material response which involves the following steps:

1 Elastic solution
2 Determination of the plastic zone
3 Mapping of the zone to a local s, t coordinate system
4 Determination of the integration scheme depending on the location of collocation point
5 Computation of the volume integral and new right hand side
6 Incremental solution with new right hand side
7 Add increments
8 Check for convergence
   If convergence achieved exit
   If not goto 2

8.1 Determination of plastic zone

Extent of zone on boundary   The explanation of how to determine the extent of the plastic zone is made for a single patch but can be extended to multiple patches. First we determine the extent of the plastic zone along the boundary using the following procedure:

- Subdivide the patch
- Starting with u = 0 and increasing u, determine for each subdivision point the state of stress[5]
- If for two subsequent points it has been found that one is elastic the other plastic then the onset of the plastic region is determined by interval halving (see Figure 6)
- The same is done for the case where the first point is plastic and the second elastic for determining the end of the plastic zone

The outcome is the local coordinate of the start of the plastic zone, u_s and its end, u_e, or in other words **a truncated NURBS curve**.

Extent of zone inside domain – predictor/corrector method Next we determine the extent of the zone inside the domain using a predictor/corrector method which is based on work first published in [4]. The aim is to determine a contour $F(\boldsymbol{\sigma}) = 0$. We can determine vectors tangential ($\mathbf{t}_c$) and normal ($\mathbf{n}_c$) to this contour:

$$
\mathbf{t}_c = \begin{pmatrix} \dfrac{\partial F}{\partial x} \\[2mm] \dfrac{\partial F}{\partial y} \end{pmatrix}; \quad \mathbf{n}_c = \begin{pmatrix} \dfrac{\partial F}{\partial y} \\[2mm] -\dfrac{\partial F}{\partial x} \end{pmatrix}
\tag{81}
$$

where

$$
\frac{\partial F(\boldsymbol{\sigma})}{\partial x} = \frac{\partial F}{\partial \boldsymbol{\sigma}} \frac{\partial \boldsymbol{\sigma}}{\partial x}; \quad \frac{\partial F(\boldsymbol{\sigma})}{\partial y} = \frac{\partial F}{\partial \boldsymbol{\sigma}} \frac{\partial \boldsymbol{\sigma}}{\partial y}
\tag{82}
$$

The derivatives of the stresses can be obtained by taking the derivatives of the equations for the internal stress computation introduced in stage 7:

$$
\frac{\partial \boldsymbol{\sigma}}{\partial x} = \sum_{e=1}^{E} \Delta \left(\frac{\partial \mathbf{S}}{\partial x} \right)^e - \sum_{e=1}^{E} \Delta \left(\frac{\partial \mathbf{R}}{\partial x} \right)^e
\tag{83}
$$

$$
\frac{\partial \boldsymbol{\sigma}}{\partial y} = \sum_{e=1}^{E} \Delta \left(\frac{\partial \mathbf{S}}{\partial y} \right)^e - \sum_{e=1}^{E} \Delta \left(\frac{\partial \mathbf{R}}{\partial y} \right)^e
$$

where E is the number of patches and the derivatives of the derived fundamental solutions $\mathbf{R}$ and $\mathbf{S}$ are given in the Appendix.

To start the iterative process we compute the inward normal to the boundary curve, $\mathbf{n}$, at the point where the plastic zone starts at the boundary. Next we make a prediction by moving along $\mathbf{n}$ with a distance h and compute the value of $F(\boldsymbol{\sigma})$ and

---

[5]Note that this calculation can be done in the local NURBS coordinate system using the stress recovery procedure presented previously.

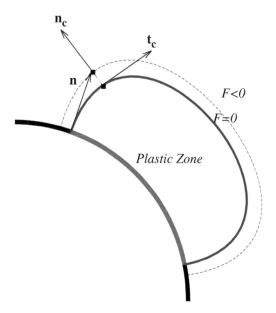

Figure 7 Explanation on how to determine the extent of the plastic zone inside the domain using the predictor/corrector method.

the vector $\mathbf{n}_c$ at this point. If, as expected $F(\boldsymbol{\sigma}) \neq 0$ then we make a correction in the direction $\mathbf{n}_c$. The new location of the point, $\mathbf{x}_{new}$, on the contour is computed by:

$$\mathbf{x}_{new} = \mathbf{x}_{old} + \mathbf{n}_c \cdot F(\boldsymbol{\sigma}) \tag{84}$$

where $\mathbf{x}_{old}$ is the point with $F(\boldsymbol{\sigma}) \neq 0$. This is shown for the case $F < 0$ in Figure 7. Since F is a nonlinear function, the location of the point computed by the simplified Equation (84) may not lie on the contour and has to be corrected. One way is to divide F into increments and update $\mathbf{n}_c$ at each increment (for more details see [4]). After the first point on the contour has been found the location of a subsequent point is predicted by moving a distance h along the tangent ($\mathbf{t}_c$) to $F(\boldsymbol{\sigma}) = 0$. This is followed by another correction.

The process continues until the end of the contour has been reached. The end of the contour can be detected, when the smallest distance of the computed point to the curve, defining the boundary of the problem, is very small. Indeed, to avoid highly singular integration it is recommended to stop at a small distance away from the boundary curve.

The method is obviously sensitive to the distance h chosen for the prediction and a compromise between computational effort and accuracy has to be reached. After all points on the contour have been determined, a NURBS curve can be constructed using an algorithm published in [6] or a polygon can be defined by connecting the points.

If the plastic domain covers the entire NURBS curve, then the process has to be modified. In this case a search is made along the inward normal at the start and the

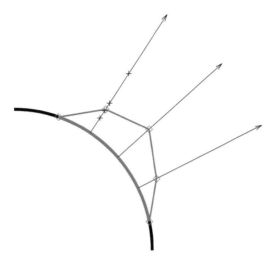

Figure 8 Explanation on how to determine the extent of the plastic zone inside the domain with the simple method. The interval halving method is indicated for the first inward normal.

end of the NURBS curve describing the boundary of the problem until the point $F = 0$ is found. The predictor/corrector method is then applied starting from the first point found and ending at the last point.

Extent of zone inside domain – simple method   A simpler alternative is presented here. For this we divide the truncated NURBS curve into suitable subdivisions. For each subdivision point we use the method of interval halving as follows:

- Compute the inward normal with a user defined length.
- Compute the yield function at the end of the line.
- If the yield function > 0 then extend the length of the line.
- If the yield function < 0 then subdivide the line and determine the yield function at the subdivision point.
- If the point is still elastic subdivide towards the boundary, if it is plastic subdivide away from the boundary.
- Do this until the elasto-plastic boundary point has been determined with sufficient accuracy.

The procedure is explained in Figure 8. We end up either with a NURBS curve or with a polygon, as shown in the Figure, that describes the elasto-plastic boundary. The accuracy of the boundary will depend on the number of subdivisions chosen on the boundary and the cut off criterion for interval halving algorithm.

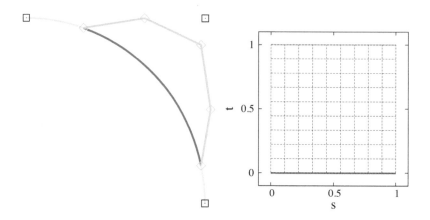

Figure 9   Explanation of the mapping of the plastic zone from the global x, y to the local s, t coordinate system. Shown are the control points of the truncated NURBS patch (red curve) and of the elasto-plastic interface approximated by a NURBS curve or order I(green curve).

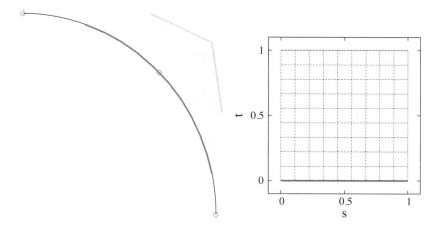

Figure 10   Mapped plastic domain in global and local coordinate system for a NURBS curve of order I. Collocation points are shown in purple.

8.2   Computation of the volume term

The computation of the volume term, as explained in section 3.2 has to be modified slightly as one of the NURBS curves is truncated and $u \neq s$. We therefore establish a relationship between local coordinate of the NURBS curve describing the boundary, u, and the coordinate s as

$$u(s) = s(u_e - u_s) + u_s \tag{85}$$

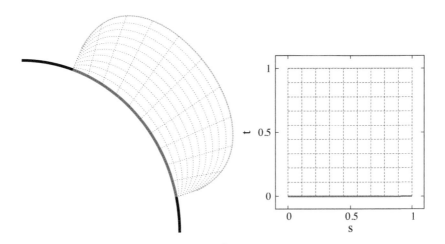

Figure 11 Mapped plastic domain in the global and the local coordinate system using a NURBS curve
of higher order.

The Jacobian of this transformation is $\frac{\partial u}{\partial s} = u_e - u_s$. The global coordinates of this
curve are then given by

$$\mathbf{x}^I(u(s)) = \sum_{n=1}^{N} R_n^I(u(s)) \cdot \mathbf{x}_n^I \tag{86}$$

where R_n^I and $\mathbf{x}_n^I$ are the basis functions and control point coordinates for the NURBS
curve describing the boundary.

For the second curve we set $u = s$ and describe it with NURBS. The simplest way
is to use a polygon i.e. a curve of order 1. If we only consider 3 points of the green
curve in Figure 9 the knot vector is $\Xi = 0, 0, 0.5, 1, 1$. The global coordinates of this
curve is then given by

$$\mathbf{x}^{II}(s) = \sum_{n=1}^{N} R_n^{II}(s) \cdot \mathbf{x}_n^{II} \tag{87}$$

where R_n^{II} and $\mathbf{x}_n^{II}$ are the basis functions and the coordinates of the control points of
the NURBS curve describing the interface.

We proceed with the mapping as explained in section 3.1. In Figure 10 we show the
resulting maps in local and global coordinate space together with the collocation points,
assuming the same basis functions are used for the approximation of the unknown.
Figure 11 shows the mapping using a higher order NURBS curve for the description
of the elasto-plastic interface.

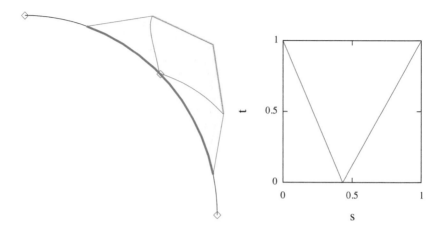

Figure 12 Subdomains for singular integration (middle collocation point) in (right) global and (left) local coordinate system.

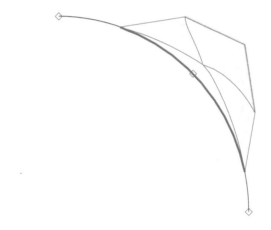

Figure 13 Subdivision of integration zone for internal stress computation.

8.3 Numerical integration

For the numerical integration we have to consider that V_0 extends to the boundary. The integrand consists of the Kernel U times the body force $\dot{\mathbf{b}}_0$ times the Jacobian. The U-Kernel is singular and approaches infinity with $O(\ln \frac{1}{r})$ as $\bar{\mathbf{x}}$ approaches the collocation point $\mathbf{y}$. When the collocation point is not part of the integration region then we can use regular integration, i.e. use Equation (20), with the number of Gauss points determined by the proximity of the collocation point. If the collocation point is part of the integration region we have to invoke singular integration as explained in stage 7, i.e. we subdivide the integration region into triangular subregions (see Figure 12).

Figure 14 Extension of plastic zone.

This means that for the integration of the volume term we will at least need several sets of Gauss points for the singular integration and for the far field integration.

8.4   Internal stress computation

To check the yield condition and to determine the residual stress and the body force term we need to compute the stresses at the Gauss points y_g. We can use Equation (44) to compute the strain and use

$$\sigma = \mathbf{C} \cdot \epsilon \tag{88}$$

to compute the stress. For the last (volume) integration in (44) it has to be considered that the Kernel S tends to infinity as $O(\frac{1}{r})$ as $\bar{x}$ approaches y_g. The integral can be regularized by a triangular subdivision as shown in Figure 13.

8.5   Extension of plastic zone during iteration

We have to consider that the plastic zone will extend during the iteration, i.e. the contour $F = 0$ will change. The simplest way would be to remap the plastic zone with a new contour.

However, this would mean that the locations of the Gauss points would change. Since we have to keep track of the stress history of points, this is not a viable option. An alternative is to add a subregion that is bounded by the previous and the new boundary. The method for the determination of the new elasto-plastic interface is similar to the one outlined for determining its initial position. A possible extension of the plastic zone is shown in Figure 14. There is the possibility that the extension of the zone is very thin, leading to problems with the numerical evaluation of the derivatives and the integral. In this case the integration can be carried out inside the original zone and an extended volume only considered if the extension is sufficiently large. Obviously a small error will be introduced here, but if an extension is skipped for a few iterations the process may be self-correcting.

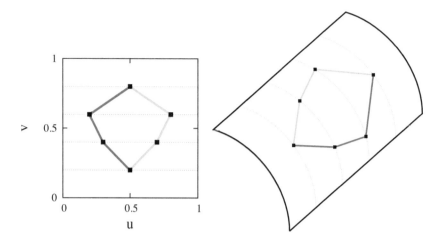

Figure 15 Determination of the trace of the plastic zone on the boundary in (left) local and (right) global coordinate space.

9   IMPLEMENTATION FOR 3-D PROBLEMS

9.1   Determination of the plastic zone

The determination of the plastic zone follows similar steps as for the plane problems. The first task is to determine the extension on the boundary.

Extent of zone on boundary   The extent of the plastic zone on the boundary involves the following steps:

- Subdivide the local coordinate space in either u or v direction.
- Scan along the lines u or v = constant[6] and find the region where $F > 0$ using the interval halving algorithm as explained for the plane case.
- Map the found points onto the global coordinate space.
- Determine 2 trimming curves that go through the points and can be used to trim the NURBS surface.

This process is demonstrated on a simple example in Figure 15 depicting the found trimming curves in green and red.

Extent of zone inside the domain – predictor/corrector method   The predictor/corrector method introduced earlier can be applied. The suggestion is to determine the inward normal, **n**, at the start of the plastic zone at a line v = const. and compute $F(\sigma)$ at a

---

[6]The decision to choose either u or v would depend on the extent of the NURBS surface, i.e. the direction in the larger dimension should be used. Scanning lines should be close enough to ensure that points are not missed. Another possibility would be to combine u and v scanning.

distance h along **n** and then correct it as done for the plane case. This proceeds until the end of the contour is found.

After this is done for all lines v = const. a point cloud is obtained through which a NURBS surface can be constructed.

Extent of zone inside the domain – simple method   To determine the extent of the plastic zone we

- Subdivide the lines u (or v) = constant into subsections and determine the inward normal at these points.
- Scan along the inward normals to determine the extent of the zone using the algorithm outlined for plane problems.
- The resulting point cloud can be used to define the surface *II* of the plastic zone.

We have now defined the two surfaces that bound the plastic zone. One is the trimmed NURBS surface describing the boundary of the problem, the other a surface defined by the point cloud.

9.2 Computation of the volume term

For the evaluation of the volume integral we proceed as outlined in section 4 but we have to consider that one of the surfaces describing the plastic zone is trimmed. We therefore need to map from the u, v coordinate system of the trimmed surface to the $\bar{u}, \bar{v}$ system as defined in section 4.1.

9.3 Numerical integration

For the volume term the integrand consists of the Kernel U times the body force $\mathbf{b}^p$ times the Jacobian. The U-Kernel is singular and approaches infinity with $O(\frac{1}{r})$ as $\bar{\mathbf{x}}$ approaches the collocation point **y**. If the collocation point is part of the integration region we have to invoke singular integration i.e. we subdivide the integration region into tetrahedal subregions in a similar way as shown for plane problems.

To check the yield condition and to determine the residual stress and the body force term we need to compute the stresses at the Gauss points **y** using Equations (44) and (88). For the last (volume) integration it has to be considered that the Kernel S tends to infinity as $O(\frac{1}{r^2})$ as **y** approaches $\bar{\mathbf{x}}$ and therefore singular integration procedures have to be used.

10 PROGRAMMING

Here we will only discuss the implementation of one aspect to volume effects, namely the effect of swelling and only for the plane strain case. The implementation of the other volume effects, such as inclusions and plasticity using NURBS technology, is subject to research at the time of writing the book, with publications expected to appear soon.

The first task is to read the geometrical information about the volume. The information is supplied in files **KnotI**, that contains knot information and **CntrlI** that contains the associated control point coordinates and weights of the curves defining the boundaries of the volume. As many volumes as required may be defined.

File **KnotI** (Knot vectors for the curves defining the volume):

```
Number of control points curve 1
Order
knot vector
Number of control points curve 2
Order
knot vector
....
```

File **CntrlI** (Control point coordinates and weights):

```
x y z w % coordinates and weight of control point 1 of curve 1
x y z w % coordinates and weight of control point 2
....
x y z w % coordinates and weight of control point 1 of curve 2
x y z w % coordinates and weight of control point 2
....
```

With this information the body force RHS is computed and added to the right hand side. Function *Bodyforce* computes the body force as follows: A call to function *Gausscor* determines the global coordinates of the Gauss points (xyg) and the Jacobian times the weight (JW).

The number of Gauss points is determined by the ratio R/L where R is the minimum distance of a collocation point to the integration region and L is the length of the region. Since Kernel **S** has the same order of singularity as Kernel **T** we use the criterion introduced earlier for selecting the number of Gauss points and the subdivision required, if the maximum number is exceeded. In reality the required number of Gauss points depends on the proximity of the collocation point and would be different for each point. Here only one criterion has been used in order to minimize the effort in computing the Jacobian at Gauss points.

```
function RHS= Bodyforce(eps0,xyp,npnts)
%-------------------
% Computes the right hand side due to initial strains
%-------------------
global E; global ny
% Determine number of Gauss locations,weights and Jacobian
[xyg,JW]= Gausscor(xyp,npnts);
RHS= zeros(npnts*2,1);
```

```
for ngp=1:columns(xyg)
 xgaus(1:2)= xyg(1:2,ngp);
 jw= JW(ngp);
 ndof=0;
 for npnt=1:npnts
  xp(1:2)= xyp(1:2,npnt);
  r= Dist(xp,xgaus);
  dxr= (xgaus-xp)/r;
  SK= SigKernel(r,dxr,E,ny,1)*jw;
  F= SK*eps0;
  ndof=ndof+1; RHS(ndof)= RHS(ndof) + F(1);
  ndof=ndof+1; RHS(ndof)= RHS(ndof) + F(2);
 end
end
endfunction
```

```
function [xyg,JW]= Gausscor(xyp,npnts)
%------------------
% Determines coordinates of Gauss points
% Input:
% xyp ... coordinates of collocation points
% npnts ... number of collocation points
%
% Output:
%  xyg  ... Gauss point coordinates
% JW ... Jacobian times weight
%------------------
fid= fopen("KnotI","r");[KnotI,nk] = fscanf(fid,"%f");fclose(fid);
fid= fopen("CntrlI","r"); CoefsI = fscanf(fid,"%f",[4,Inf]);fclose(fid);
[ngausu,ngausv,nsubu,nsubv]= GpointsI(KnotI,CoefsI,xyp,npnts);
[su,us]= SubdivI(nsubu); [sv,vs]= SubdivI(nsubv);
[Coru,Wiu]= Gauss(ngausu); [Corv,Wiv]= Gauss(ngausv);
%----------------------------
%  Gauss integration of volume term
%----------------------------
Jacuv= su*sv;
ngp=0;
for ndv=1:nsubv
 for ndu=1:nsubu
  for ngv=1:ngausv
   for ngu=1:ngausu
    s(ngu)= su/2*(Coru(ngu)+1)+us(ndu);
    t(ngu)= sv/2*(Corv(ngv)+1)+vs(ndv);
   end
  end
  [xy,J]= Map(KnotI,CoefsI,s,t);
```

```
  for ngv=1:ngausv
   for ngu=1:ngausu
    ngp=ngp+1;
    xyg(:,ngp)= xy(:,ngu,ngv);
    JW(ngp)= Wiu(ngu)*Wiv(ngv)*J(ngu,ngv)*Jacuv;
   end
  end
 end
end
endfunction
```

```
function [ngausu,ngausv,nsubu,nsubv]= GpointsI(KnotI,CoefsI,xyp,npnts)
%--------------------
% Determines number of Gauss points for volume integration
% Input:
%  KnotI  ... knot vectors of bounding curves
%  CoefsI ... Control point coords and weights
%  xyp  ... collocation point coords.
%  npnts ... numer of colloctaion points
%
% Output:
% ngausu,ngausv  ...  Number of Gauss points in s,t directions
% nsubu,nsubv  ...  Number of subdivisions in s,t directions
%--------------------
ndiv=5; s= linspace(0,1,ndiv); t=linspace(0,1,2); Rmin=1000;
xy= Map(KnotI,CoefsI,s,t);
for n=1:length(s)
 xyn(1:2)= xy(:,n,1);
 for np=1:npnts
  r= Dist(xyn,xyp(:,np));
  if(r < Rmin) Rmin=r; endif
 end
end
Lx= Dist(xy(:,1,1),xy(:,ndiv,1)); Ly= Dist(xy(:,1,1),xy(:,1,2));
[ngausu,nsubu]= Gausspoints(Rmin,Lx,1);
[ngausv,nsubv]= Gausspoints(Rmin,Ly,1);
endfunction
```

```
function SK  = SigKernel(r,dxr,E,ny,nstress)
% --------------------------
% Computes Stress Kernel for plane strain
%
% INPUT:
% r ...  distance betewen source and field point
% dxr ...   derivatives of r
```

```
% E   ... modulus of elasticity
% ny ... Poissons ratio
% nstress  ... stress state indicator
% (1= plane strain, 2= plane stress)
%
% OUTPUT:
% SK    ... Stress Kernel
%-----------------------
if(nstress == 1)
 c2= 1.0/(4.0*pi*(1.0-ny));c3= 1.0-2.0*ny; c4= 2.0; c5=1;
else
 c2= (1.0-ny)/4.0*pi; c3= (1.0-ny)/(1+ny); c4= 2.0; c5=(1-ny)/(1+ny);
endif
SK(1,1)= c3*2*dxr(1) - c5*dxr(1) + c4*dxr(1)**3;
SK(2,2)= c3*2*dxr(2) - c5*dxr(2) + c4*dxr(2)**3;
SK(1,2)= -c5*dxr(1) + c4*dxr(1)*dxr(2)*dxr(2);
SK(2,1)= -c5*dxr(1) + c4*dxr(2)*dxr(1)*dxr(1);
SK(1,3)= c3*dxr(2) + c4*dxr(1)*dxr(1)*dxr(2);
SK(2,3)= c3*dxr(1) + c4*dxr(2)*dxr(1)*dxr(2);
SK= -SK*c2/r;
endfunction;
```

II EXAMPLE

As an example we show a circular hole in an infinite domain subjected to a volume increase in an area above it. The same assumptions and material properties as for the example of stage 7 are used. The volume change is simulated by applying an initial strain in vertical direction of 10% above the hole. The additional input data required are:

```
KnotI:
2
1
0 0 1 1
2
1
0 0 1 1

CntrlI:
-1.5 1.5 0 1
1.5 1.5 0 1

-1.5 2 0 1
1.5 2 0 1
```

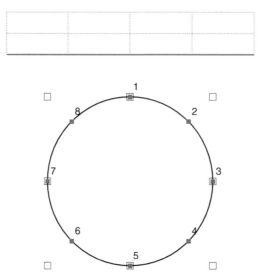

Figure 16 Circular hole subject to swelling: Geometry definition with top and bottom curves defining the zone of swelling marked in red and green. Control points for the definition of the circle are marked by hollow squares. Red squares indicate collocation points before refinement.

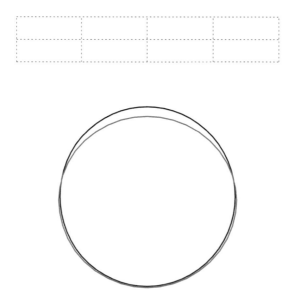

Figure 17 Circular hole subject to swelling: Displaced shape.

Figure 16 shows the geometry definition as well as the collocation points before refinement. The result of the simulation, namely the deflected shape of the hole is shown in Figure 17.

12   SUMMARY AND CONCLUSIONS

We have introduced volume effects into our BEM formulation. This allows the treatment of heterogeneous domains and inelastic material behavior, an important aspect for the method to be applicable to practical problems. This is an area of research that in the opinion of the author has been neglected. Published work on volume effects date a few years back and are not user friendly. A method still prevalent is the use of cells for the volume integration. This not only means an additional discretization effort but also a loss of accuracy because continuous basis functions are used for the approximation of body forces inside cells. Considering that the elasto-plastic boundary will transect some cells this is an unacceptable source of error.

The reader may observe a paucity of examples, that demonstrate that the theory is working. This is because some of the ideas proposed here of applying NURBS technology are quite novel and subject to research. Indeed the aim of the chapter is to present these new ideas, so that a larger number of researchers is encouraged to work in this area. The approaches presented here have the advantage that no discretization is required, that the plastic zone is determined automatically and that integration is carried out inside the plastic zone. Note that this approach is more accurate than some implementations used in the FEM with continuous basis functions, where finite elements may be partially elastic and partially plastic.

It was shown that NURBS technology has simplified the definition of the geometry of inclusions, the determination of the plastic zone and the volume integration.

BIBLIOGRAPHY

[1]   I.C. Cormeau. Numerical stability in quasi-static elasto-viscoplasticity. *International Journal for Numerical Methods in Engineering*, 9(1), 1975.

[2]   X.-W. Gao and T. G. Davies. *Boundary Element Programming in Mechanics*. Cambridge University Press, Cambridge, 2002.

[3]   R. Hill. *The mathematical theory of plasticity*. Oxford University press, 1950.

[4]   M. Noronha, A.S. Müller, and A. M. B. Pereira. A novel pure-BEM approach for post-processing and non-linear analysis. In *Proceedings of MacMat2005, Louisiana, USA*, 2005.

[5]   G. Pande, G. Beer, and J. Williams. *Numerical methods for rock mechanics*. J. Wiley, 1990.

[6]   Les Piegl and Wayne Tiller. *The NURBS book (2nd ed.)*. Springer-Verlag New York, Inc., New York, NY, USA, 1997.

[7]   W. Prager and P.G. Hodge. *Theory of perfectly plastic solids*. J. Wiley, 1951.

[8]   T. Ribeiro, G. Beer, and C. Duenser. Efficient elastoplastic analysis with the boundary element method. *Computational Mechanics*, 41:715–732, 2008.

[9]   K. Riederer, C. Duenser, and G. Beer. Simulation of linear inclusions with the BEM. *Engineering Analysis with Boundary Elements*, 33(7):959–965, 2009.

[10]   I. M. Smith, D. V. Griffiths, and L. Margetts. *Programming the Finite Element Method*. Wiley, 2013.

[11]   J.C.F. Telles. *The boundary element method applied to inelastic problems*. Springer, 1983.

Chapter 11

Stage 10: The time domain

Time is an illusion

A. Einstein

where we consider a 4th dimension (time).

1 INTRODUCTION

In all our deliberations so far we have neglected the dimension time, i.e. we have assumed that the domain does not have any mass or that the loading and response to the loading is instantaneous, so time does not play any role. Here we introduce time and – more importantly – mass effects. The following is only a short introduction. More details can be found, for example, in [1].

1.1 Bernoulli beam with mass

Consider the simply supported beam in Figure 1, which now is assumed to have mass. If this is the case, then additional forces occur, the main one being the resistance of the mass to acceleration (inertia effects). There are two inertia effects: one is the resistance against translation, the other against rotation.

According to Newton's law the translational inertia force is given by:

$$F_m(x) = m \cdot dx \cdot \ddot{w}(x) \tag{1}$$

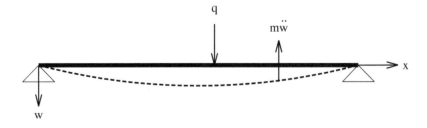

Figure 1 Simply supported beam with mass.

Figure 2 Deflected shape of the beam at different times.

where F is the inertia force (kN), m is the mass per unit length (kg/m) and $\ddot{w}(x)$ is the downward acceleration (m/sec^2) at point x. The distributed inertia force is $f = \frac{F}{dx}$.

For the beam the effect of rotational inertia is small, so it will be neglected here. An additional force occurs due to damping. This force depends on the velocity of the deflection ($\dot{w}(x)$):

$$F_d(x) = c \cdot dx \cdot \dot{w}(x) \tag{2}$$

where c is a damping coefficient.

Substitution of this into the differential equation for beam bending derived earlier gives

$$\frac{d^2}{dx^2}(EIw'') + q - m \cdot \ddot{w} - c \cdot \dot{w} = 0 \tag{3}$$

2 SOLUTIONS IN THE FREQUENCY DOMAIN

We can set the loading q to zero and obtain a solution with the inertia loading only. This is equivalent to applying a deformation to the beam and letting it vibrate freely (similar to plucking the string of a guitar). In this case the vibration will be harmonic, i.e. the deflection is given by

$$w(x,t) = w(x,0) \cdot \sin(\omega t) \tag{4}$$

where $w(x,0)$ is the displacement amplitude at location x, ω is the angular frequency and t is the time. Figure 2 shows the deflected shape of the beam at 4 different times.

The second derivative of w is given by

$$\ddot{w}(x,t) = \omega^2 \cdot w(x,0) \cdot \sin(\omega t) \tag{5}$$

There are (for this case) an infinite number of frequencies that the beam can vibrate and these are called *Eigen frequencies* or *natural frequencies*. The relationship between angular frequency and frequency, measured in cycles per second (Hertz) is

$$f = \frac{\omega}{2\pi} \tag{6}$$

We further define as period T, the time it takes to complete a cycle:

$$T = \frac{1}{f} \tag{7}$$

Figure 3 First 3 Eigenmodes.

For each frequency there is an Eigen-mode, i.e. the form in which the beam vibrates. Figure 3 shows the first 3 Eigen-modes.

Setting $q = 0$ and neglecting (for the moment) the damping force we substitute the expression 4 and 5 into Equation 17 and obtain

$$\frac{d^2}{dx^2}(EIw'') \cdot \sin(\omega t) - m \cdot \omega^2 \cdot w \cdot \sin(\omega t) = 0 \tag{8}$$

or after multiplying the equation with $\frac{1}{\sin(\omega t)}$

$$\frac{d^2}{dx^2}(EIw'') - m \cdot \omega^2 \cdot w = 0 \tag{9}$$

2.1 Numerical solution

For the numerical solution we use the principle of virtual work as explained previously. The external virtual work done by the inertia force is

$$\delta W_e^d = \int_{x=0}^{L} \omega^2 \cdot m \cdot \tilde{w} \cdot \delta w \cdot dx \tag{10}$$

After discretization we obtain for the virtual work (assuming m is constant):

$$\delta W_e^d = \int_{-1}^{1} \omega^2 \cdot m \cdot \left(\sum_{i=1}^{I} N_i(\xi) \cdot w_i \right) \cdot N_j(\xi) \cdot \delta w_j \cdot J \cdot d\xi$$

$$= \omega^2 \cdot m \cdot \left[\sum_{i=1}^{I} \left(\int_{-1}^{1} N_i(\xi) \cdot N_j(\xi) \cdot J \cdot d\xi \right) \cdot w_i \right] \cdot \delta w_j \tag{11}$$

Following the procedures in Stage 5, we obtain the following equation for the coefficients of the stiffness matrix:

$$K_{ij}^d = \int_{-1}^{1} \left[E \cdot I \cdot \frac{\partial^2 N_i(\xi)}{\partial x^2} \cdot \frac{\partial^2 N_j(\xi)}{\partial x^2} - \omega^2 \cdot m \cdot N_i(\xi) \cdot N_j(\xi) \right] \cdot J \cdot d\xi \tag{12}$$

This is also known as the *dynamic stiffness* matrix, which can be split into the static stiffness matrix and mass matrix. The system of equations can be written in matrix form as

$$\left(\mathbf{K} - \omega_n^2 \mathbf{M} \right) \mathbf{w} = 0 \tag{13}$$

where we have introduced the subscript n for the n-th Eigenfrequency. $\mathbf{K}$ is the static stiffness matrix derived earlier and $\mathbf{M}$ is the mass matrix, whose coefficients are defined as

$$M_{ij} = m \cdot \int_{-1}^{1} N_i(\xi) \cdot N_j(\xi) \cdot J \cdot d\xi \tag{14}$$

The number of Eigenfrequencies that can be obtained depends on the number of degrees of freedom, i.e. the size of $\mathbf{K}$.

Lumped mass matrix   The mass matrix just derived is also known as a *consistent mass matrix*. Another simpler alternative is often used where the mass is assumed to be concentrated at various points along the beam and is also known as *Lumped mass matrix*. It is a diagonal matrix, where the coefficients are defined by:

$$M_{jj} = m \cdot \int_{-1}^{1} N_j(\xi) \cdot J \cdot d\xi \tag{15}$$

Eigensolution   Since the right hand side is zero, equation (13) can not be solved for the unknown displacements $\mathbf{w}$. For the non-trivial case where the displacements $\mathbf{w}$ are not zero, the determinant of the expression in parentheses must be zero, i.e.:

$$|\mathbf{K} - \omega_n^2 \mathbf{M}| = 0 \tag{16}$$

which allows us to determine the Eigenfrequencies ω_n, that coincide with locations where the determinant becomes zero. After determining an Eigenfrequency we may compute the *Eigenforms* or *Eigenvectors* by setting one coefficient of vector $\mathbf{w}$ to one. If this is done the column of $\mathbf{K}^d$, associated with the known value can be transferred to the right hand side and the system can be solved for the remaining coefficients of $\mathbf{w}$. The Eigenvector therefore constitutes the displacements relative to the defined value, i.e. the shape.

Programming   The changes to the program introduced in Stage 5 are relatively minor. We compute a consistent mass matrix which readers may recognize as being similar to the matrix for the elastic foundation with $-\omega_n^2$ replacing the spring stiffness k_w or a lumped mass matrix, which is similar to determining the nodal point forces with $-m$ instead of q.

The Eigenvectors and frequencies can be computed using an octave intrinsic function:

```
[Eigenvect,omega]= eig(K,M)
```

3 SOLUTIONS IN THE TIME DOMAIN

Here we discuss solutions where the loading is a function of time. We start with the differential equation of a system with one degree of freedom (mass suspended on a spring):

$$m \cdot \ddot{w} + c \cdot \dot{w} + k \cdot w = q(t) \qquad (17)$$

where $q(t)$ is function of time, k is the spring stiffness and m and c are mass and damping coefficient.

For a general variation of q with time the differential equation can only be solved numerically. There are two approaches for the solution, the finite difference and the *Newmark* method. Both methods rely on a discretization of the time domain, i.e. we divide the time into equal time steps Δt.

3.1 Finite difference method

Here we approximate the time derivatives with a central difference. The first derivative of w is given by (see Figure 4):

$$\dot{w}_i = \frac{w_{i+1} - w_{i-1}}{\Delta t} \qquad (18)$$

The second derivative is:

$$\ddot{w}_i = \frac{\frac{w_{i+1} - w_i}{\Delta t} - \frac{w_i - w_{i-1}}{\Delta t}}{\Delta t} = \frac{w_{i+1} - 2w_i + w_{i+1}}{\Delta t^2} \qquad (19)$$

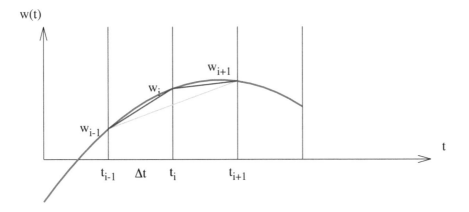

w(t)

w_{i+1}

w_i

w_{i-1}

t_{i-1} Δt t_i t_{i+1} t

Figure 4 Explanation of the finite difference method. The approximation of the velocity is depicted in green, the one for the acceleration in blue.

The equilibrium equation (17) can now be written for time t_i as:

$$m \cdot \ddot{w}_i + c \cdot \dot{w}_i + k \cdot w_i = q_i \qquad (20)$$

Substitution of the finite difference expressions results in:

$$m \left[\frac{w_{i+1} - 2w_i + w_{i-1}}{\Delta t^2} \right] + c \left[\frac{w_{i+1} - w_{i-1}}{\Delta t} \right] + kw_i = q_i \qquad (21)$$

or putting the unknown value on the left hand side

$$\left[\frac{m}{\Delta t^2} + \frac{c}{2\Delta t} \right] w_{i+1} = q_i - \left[\frac{m}{\Delta t^2} + \frac{c}{2\Delta t} \right] w_{i-1} - \left[k - \frac{2m}{\Delta t} \right] w_i \qquad (22)$$

For the solution we need two initial values of displacements: w_i, w_{i-1}, at the start of time (0), namely w_0, w_{-1}. w_{-1} can be obtained by defining initial values for $\dot{w}_0$ and $\ddot{w}_0$. The finite difference expression:

$$\dot{w}_0 = \frac{w_1 - w_{-1}}{2\Delta t}; \quad \ddot{w}_0 = \frac{w_1 - 2w_0 + w_{-1}}{2\Delta t^2} \qquad (23)$$

can be solved for w_{-1}:

$$w_{-1} = w_0 - \Delta t \cdot \dot{w}_0 + \frac{\Delta t^2}{2} \ddot{w}_0 \qquad (24)$$

The method only gives a stable solution for

$$\frac{\Delta t}{T_n} < \frac{1}{\pi} \qquad (25)$$

where T_n is the first Eigenperiod of the system.

3.2 Newmark method

There is some scope for improvement. Newmark[1] had the idea that if $\ddot{w}$ is approximated instead of w, then there would be a considerable improvement of the approximation of w, since it is obtained by a double integration in time of the acceleration. Figure 6 shows an example of a possible variation of w, $\dot{w}$ and $\ddot{w}$ with time. The basic idea of Newmark is to assume a variation of $\ddot{w}$ inside the time step Δt.

If this is done we can express the variation of the velocity and deformation by a repeated integration of the acceleration. We may assume, for example, that the acceleration remains constant during the time step, which gives the following result

---

[1] Nathan Mortimore Newmark (September 22, 1910 to January 25, 1981) was an American structural engineer and academic, who is widely considered as one of the founding fathers of Earthquake Engineering.

Figure 5 Nathan Newmark.

for the time $t_i + \tau$, where τ is measured from t_i:

$$\ddot{w}(t_i + \tau) = \ddot{w}_i + \delta(\ddot{w}_i + \ddot{w}_{i+1}) \tag{26}$$

or

$$\ddot{w}(t_i + \tau) = (1 - \delta)\ddot{w}_i + \delta\ddot{w}_{i+1}$$

where $0 \leqslant \delta \leqslant 1$ is a constant that will be specified later. Using this assumption the velocity can be computed by:

$$\dot{w}(t_i + \tau) = \dot{w}_i + \int_0^\tau \ddot{w}(\tau)d\tau = \dot{w}_i + [(1 - \delta)\ddot{w}_i + \delta\ddot{w}_{i+1}]\,\tau \tag{27}$$

The deformation w can be obtained by a further integration:

$$w(t_i + \tau) = w_i + \int_0^\tau \dot{w}(\tau)d\tau = w_i + \tau \cdot \dot{w}_i + \frac{\tau^2}{2}[(1 - \delta)\ddot{w}_i + \delta\ddot{w}_{i+1}] \tag{28}$$

The Newmark formula can be obtained by setting $\beta = \frac{\delta}{2}$:

$$w(t) = w_i + \dot{w}_i \cdot \tau + \left[\left(\frac{1}{2} - \beta\right)\ddot{w}_i + \beta \cdot \ddot{w}_{i+1}\right]\tau^2 \tag{29}$$

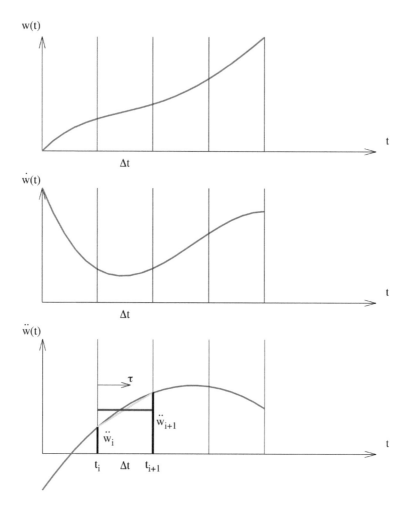

Figure 6 Variation of displacement, velocity and acceleration with time, showing possible approximation of acceleration.

where the total time $t = t_i + \tau$ has been introduced. For an average acceleration we set $\delta = \frac{1}{2}$ $(\beta = \frac{1}{4})$ and obtain

$$w(t) = w_i + \dot{w}_i \cdot \tau + [\ddot{w}_i + \ddot{w}_{i+1}] \frac{\tau^2}{4} \qquad (30)$$

At time $t_i + \Delta t$ the displacement w_{i+1} can be computed by:

$$w_{i+1} = w_i + \dot{w}_i \cdot \Delta t + [\ddot{w}_i + \ddot{w}_{i+1}] \frac{\Delta t^2}{4} \qquad (31)$$

The acceleration is:

$$\ddot{w}_{i+1} = \frac{4}{\Delta t^2} \left[w_{i+1} - w_i - \dot{w}_i \Delta t - \ddot{w}_i \frac{\Delta t^2}{4} \right] \qquad (32)$$

The velocity at time t_{i+1} is given by:

$$\dot{w}_{i+1} = \dot{w}_i + \frac{1}{2}(\ddot{w}_i + \ddot{w}_{i+1})\Delta t \tag{33}$$

and after substitution of (32) we get:

$$\dot{w}_{i+1} = -\dot{w}_i + \frac{2}{\Delta t}(w_{i+1} - w_i) \tag{34}$$

The equations of equilibrium can be written for time t_{i+1} as:

$$m\ddot{w}_{i+1} + c\dot{w}_{i+1} + kw_{i+1} = q_{i+1} \tag{35}$$

or after the substitution of the Newmark formula and rearranging terms:

$$\left[\frac{4m}{\Delta t^2} + \frac{2c}{\Delta t} + k\right] w_{i+1} = q_{i+1} + \frac{m}{\Delta t^2}\left[4w_i + 4\dot{w}\Delta t + \ddot{w}_i\Delta t^2\right] + \frac{c}{\Delta t}\left[2w_i + w_i\Delta t\right] \tag{36}$$

where the dynamic stiffness matrix is given by:

$$k^d = k + \frac{2}{\Delta t} \cdot c + \frac{4}{\Delta t^2} \cdot m \tag{37}$$

The right hand side for time step $i + 1$ is defined by:

$$P_{i+1} = q_{i+1} + \frac{m}{\Delta t^2}\left[4w_i + 4\dot{w}\Delta t + \ddot{w}_i\Delta t^2\right] + \frac{c}{\Delta t}\left[2w_i + w_i\Delta t\right] \tag{38}$$

For constant acceleration the method is stable for any value of Δt. The Newmark method allows assumptions other than average acceleration. By putting $\beta = \frac{1}{6}$ for example, a linear variation is assumed within the time step. Note that for the assumption of a constant acceleration (order 0), the approximation for the velocity and the displacement is of order 1 and 2 respectively. The assumption for the load q is however, that it is constant within the time step. For a system with more than one degree of freedom we have:

$$\mathbf{K}^d \mathbf{u}_{i+1} = \mathbf{P}_{i+1} \tag{39}$$

with

$$\mathbf{K}^d = \mathbf{K} + b_3 \cdot \mathbf{C} + b_1 \cdot \mathbf{M} \tag{40}$$

where $\mathbf{K}$, $\mathbf{C}$ and $\mathbf{M}$ is the static stiffness, damping and mass matrix respectively and

$$\mathbf{P}_{i+1} = \mathbf{F}_{i+1} + \mathbf{M}\left[b_1\mathbf{u}_i + b_2\dot{\mathbf{u}} + \ddot{\mathbf{u}}_i\right] + \mathbf{C}\left[b_3\mathbf{u}_i + \mathbf{u}_i\right] \tag{41}$$

where $\mathbf{F}_{i+1}$ is the load vector at time $i + 1$ and

$$b_1 = \frac{4}{\Delta t^2}; \quad b_2 = \frac{4}{\Delta t}; \quad b_3 = \frac{2}{\Delta t} \tag{42}$$

4 PROGRAMMING

We implement dynamic capabilities into the beam program. A program flow chart is given below:

- INITIAL CALCULATIONS
  Form $\mathbf{K}$, $\mathbf{M}$ and $\mathbf{C}$
  Specify time step $\triangle t$ and compute constants b_1 to b_3
  Calculate $\mathbf{K}^d$ using (40)
  Specify initial conditions w_1, $\dot{w}_1$, $\ddot{w}_1$
- FOR $i = 1$ to number of time steps
  Calculate $\mathbf{P}_{i+1}$ using (41)
  Solve for displacements $\mathbf{w}_{i+1} = \left(\mathbf{K}^d\right)^{-1}\mathbf{P}_{i+1}$
  Calculate $\dot{w}_{i+1}$ using (33)
  and $\ddot{w}_{i+1}$ using (32)
- END DO

The method is implemented for the Bernoulli beam of stage 4. The function *BernoulliNewmark* considers a simply supported beam with mass and the following properties

- $EI = 1/100\,\mathrm{Nm}^2$, $\nu = 0$
- $m = 1\,\mathrm{kg/m}$, $c = 0$ (no damping)
- $q = 1\,\mathrm{N/m}$ applied at time 0 and removed at time Timp
- $L = 1\,\mathrm{m}$

```
function BernoulliNewmark
%------------------------------------
%  simply supported beam, Newmark method
%  impact load q, no damping
%------------------------------------
global I; global p; global EI; global Jac; global mult; global smass
fid= fopen("Thist","w");
%  Input data:
EI=1/100; L=1; smass= 1;  q=1;
Deltat=0.1; Jac= L; mult=(1/Jac)^2;
Timp=6.4;    % time when load is removed
%  constants:
b1= 4/(Deltat^2); b2= 4/Deltat; b3=2/Deltat;
b4= Deltat^2/4;
p=4; ki=0;
% compute dynamic stiffness matrix and mass matrix
[K,M,F]= BernoulliKM(p,ki,b1,q);
[ncu,knot]= Knotok(p,ki);
%  initial conditions:
for n=1:ncu-2
 w(n,1)= 0; v(n,1)= 0; a(n,1)= 0;
end
wi=w; vi=v; ai=a;
Ntimes= 200; Time=0.0;
```

```
fprintf(fid,"%f %f \n", 0, 0)
for nt=1:Ntimes
 Time= Time + Deltat;
 if(nt == 1)
  P= F;
 elseif(Time <= Timp)
  P= F + M*(b1*w + b2*v + a);
 else
  P= M*(b1*w + b2*v + a);
 endif
 w= K\P; v= -vi + b3*(w - wi);
 a= b1*(w - wi -vi*Deltat - ai*b4);
 wi=w; vi=v; ai=a;
 ut=0.5
 disp=Getval(ncu,knot,w,ut);
   fprintf(fid,"%f %f \n", Time, disp(1))
end
fclose(fid);
endfunction;
```

```
function [K,Mass,F]= BernoulliKM(p,ki,b1,q)
%-----------------------------------
%  Computes the dynamic stiffness matrix
%  mass matrix and force vector
%  for Bernoulli beam with Newmark
%  Input:
%  p,ki ... order of basis functions and number of inserted knots
%  b1  ...   multiplier for Mass matrix
%  q  ... distributed load
%
% Output:
%  K  ... dynamic stiffness matrix
%  Mass  ... mass matrix
%  F  ... force vector
%-----------------------------------
global EI; global Jac; global smass
[I,knot,nsub,us]= Knotok(p,ki);
M=8; [xsig,W]= Gauss(M);
[ug,Jacu,Mg,Wg]= Getu(M,xsig,W,nsub,us);
[N,d2N]=Getfun(knot,ug,I,p);
%   compute dynamic stiffness matrix and mass matrix
for i=1:I
  for j=1:I
    kij= 0; mij= 0;
    for m=1:Mg
     kij= kij + EI*d2N(i,m)*d2N(j,m)*Jac*Jacu(m)*Wg(m);
     mij= mij + smass*N(i,m)*N(j,m)*Jac*Jacu(m)*Wg(m);
    end
    k(i,j)= kij + b1*mij;
```

```
      mm(i,j)=mij;
  end
end
for i=1:I
 f(i)= 0;
 for m=1:Mg
  f(i)= f(i) + q*N(i,m)*Jac*Jacu(m)*Wg(m);
 end
end
%  assign boundary conditions (delete rows, columns)
ii=0;
for i=2:I-1
 ii=ii+1; F(ii,1)= f(i);
 jj=0;
 for j=2:I-1
 jj=jj+1;
 K(ii,jj)= k(i,j);
 Mass(ii,jj)= mm(i,j);
 end
end
endfunction;
```

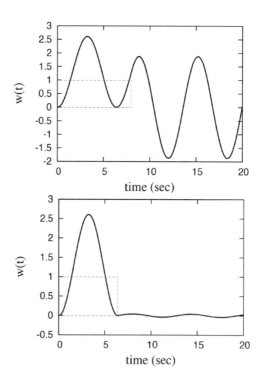

Figure 7 Beam subjected to sudden load: Variation of displacement with time. Duration of loading indicated in red. Top: Timp = 8 sec. Bottom: Timp = 6.4 sec.

A good approximate solution for the beam can be obtained with a B-spline of order $p = 4$ and no knot insertions. The first Eigenperiod of the beam T_n is 6.366 sec. The response of the beam depends on the time Timp when the load is removed and is related to the Eigenperiod. In Figure 7 the variation of the displacement is shown with time for the case Timp $= 6.4$ sec and Timp $= 8$ sec. They compare well with dynamic responses published in [1].

5   SUMMARY AND CONCLUSIONS

For this stage the time dimension was introduced in the analysis, i.e. we have considered mass and damping effects. For the simulation in the time domain, we introduced two numerical methods: The Finite Difference and the Newmark method. The latter is the preferred solution method and as has been shown in the example of the beam, results in an accurate simulation of the dynamic response.

BIBLIOGRAPHY

[1]   A. K. Chopra. *Dynamics of structures*. Prentice Hall, 2011.

Appendix: Fundamental solutions

Here functions are presented for some of the fundamental solutions used in the book based on solutions published in [1], which also contains a complete set. The derivatives of **S** and **R** have been published in [2]. The non-singular part of the stress and strain solutions is first determined in tensor notation and then converted to Voight notation. The functions use the distance between field point and source point, r and its first and second derivatives, which are computed using function Rderivs.

```
function [r,dxr,dxr2]= Rderivs(x,y)
% -------------------------
% Computes r and first and second derivatives
% Input:
% x ... coordinates of field point
% y ... coordinates of source point
% Output:
% r   ... distance between x and y
% dxr   ... first derivatives
% dxr2 ...   second derivatives
%----------------------
dx2=0; ndim=length(x);
for i=1:ndim
 dx(i)= x(i) - y(i);
 dx2= dx2 + dx(i)^2;
end
r= sqrt(dx2);
if(nargout == 1) return endif
for i=1:ndim
  dxr(i)= dx(i)/r;
end
if(nargout == 2) return endif
for i=1:ndim
 for j=1:ndim
   if(i == j)
    dxr2(i,j)= 1/r - (x(i) - y(i))^2/r^3;
   else
    dxr2(i,j)= - (x(i) - y(i))*(x(j) - y(j))/r^3;
   endif
```

```
 end
end
endfunction;
```

I STRESS SOLUTION Σ(x,y)

```
function S= Sijk(i,j,k,c3,c4,dxr)
% -------------------------
% Computes nonsingular part of derived Kernel
% for internal stress solution (to be multiplied with t)
%
% INPUT:
% i,j,k  ...  indices
% c3,c4  ... constants
% dxr ... r,i
%
% OUTPUT:
% S   ... derived Kernel
%-----------------------
S=0;
if(i == k) S=S+c3*dxr(j); endif
if(j == k) S=S+c3*dxr(i); endif
if(i == j) S=S-c3*dxr(k); endif
S=S+c4*dxr(i)*dxr(j)*dxr(k);
endfunction;

function SK  = SKernel(r,dxr,ny,nstress)
% -------------------------
% Computes derived Kernel
% for internal stress solution (to be multiplied with t)
%
% INPUT:
% r ...  distance between source and field point
% dxr ...   first derivates of r
% ny ... Poissons ratio
% nstress  ... stress state indicator
% (1= plane strain, 2= plane stress, 3= 3D)
%
% OUTPUT:
% SK   ... derived Kernel
%-----------------------
ndim=2; nstr=3;
if(nstress == 1)
 c= 1.0/(4.0*pi*(1.0-ny)*r);c3= 1.0-2.0*ny; c4= 2;
elseif(nstress == 2)
 c= (1.0+ny)/(4.0*pi*r); c3= (1.0-ny)/(1+ny); c4= 2;
```

```
elseif(nstress == 3)
 c= 1.0/(8.0*pi*(1.0-ny)*r^2); c3= 1.0-2*ny; c4= 3;
 ndim= 3; nstr=6;
endif
%  Voight notation
if( nstress < 3)
 I(1:3)= [1 2 1];
 J(1:3)= [1 2 2];
else
 I(1:6)= [1 2 3 1 2 1];
 J(1:6)= [1 2 3 2 3 3];
endif
for k=1:ndim
 for ns=1:nstr
  i=I(ns); j=J(ns);
  SK(ns,k)= Sijk(i,j,k,c3,c4,dxr);
 end
end
SK= SK*c
endfunction;
```

2 DERIVED SOLUTION FOR DISPLACEMENT $\bar{S}(x,y)$

```
function S= Sbarijk(i,j,k,c3,c4,dxr)
% --------------------------
% Computes nonsingular part of derived Kernel
% for internal strain solution (to be multiplied with t)
%
% INPUT:
% i,j,k  ... indices
% c3,c4  ... constants
% dxr ... first derivatives of r
%
% OUTPUT:
% S   ... derived Kernel
%-----------------------
S=0;
if(i == k) S=S+c3*dxr(j); endif
if(j == k) S=S+c3*dxr(i); endif
if(i == j) S=S-c3*dxr(k); endif
S=S+c4*dxr(i)*dxr(j)*dxr(k);
endfunction;

function SKbar  = SbarKernel(r,dxr,E,ny,nstress)
% --------------------------
```

```
% Computes derived Kernel
% for internal strain solution (to be multiplied with t)
%
% INPUT:
% r ...   distance between source and field point
% dxr ...   first derivatives of r
% E  ... modulus of elasticity
% ny ... Poissons ratio
% nstress  ... stress state indicator
% (1= plane strain, 2= plane stress, 3= 3D)
%
% OUTPUT:
% SK   ... derived Kernel
%-----------------------
ndim=2; nstr=3; G= E/(2*(1+ny));
if(nstress == 1)
 c3= 1.0-2.0*ny; c4= 2; c=1/(8*pi*G*(1-ny)*r);
elseif(nstress == 2)
  c3= (1.0-ny)/(1+ny); c4= 2; c=(1+ny)/(8*pi*G*r);
elseif(nstress == 3)
  c3=1.0-2*ny; c4=3; c=1/(16*pi*G*(1-ny)*r^2); ndim=3; nstr=6;
endif
%  Voight notation
if( nstress < 3)
 I(1:3)= [1 2 1];
 J(1:3)= [1 2 2];
else
 I(1:6)= [1 2 3 1 2 1];
 J(1:6)= [1 2 3 2 3 3];
endif
for k=1:ndim
 for ns=1:nstr
  i=I(ns); j=J(ns);
  SKbar(ns,k)= Sbarijk(i,j,k,c3,c4,dxr);
 end
end
SKbar= SKbar*c
endfunction;
```

3 DERIVED SOLUTION FOR TRACTION $\bar{R}(x,y)$

```
function R= Rbarijk(i,j,k,c3,c4,c6,ny,dxr,vnor)
% -------------------------
% Computes nonsingular part of derived Kernel
% for internal strain solution (to be multiplied with u)
```

```
%
% INPUT:
% i,j,k  ...  indices
% c3,c4,c6,c7,ny  ... constants
% dxr ... first derivatives of r
% vnor  ... outward normal
%
% OUTPUT:
% R   ... derived Kernel
%-----------------------
cost= vecdotp(vnor,dxr); mult= c4*cost;
R=0;
if(i == k) R= R + ny*dxr(j); endif
if(j == k) R= R + ny*dxr(i); endif
if(i == j) R= R * dxr(k); endif
R= R - c6*dxr(i)*dxr(j)*dxr(k);
R= mult*R;
if(i == k) R= R + c3*vnor(j); endif
if(i == j) R= R - c3*vnor(k); endif
if(j == k) R= R + c3*vnor(i); endif
R= R + c3*c4*dxr(i)*dxr(j)*vnor(k);
R=R + c4*ny*(vnor(j)*dxr(i)*dxr(k) + vnor(i)*dxr(j)*dxr(k));
endfunction;


function RKbar  = RbarKernel(r,dxr,vnor,E,ny,nstress)
% -------------------------
% Computes derived Kernel
% for internal strain solution (to be multiplied with u)
%
% INPUT:
% r ...  distance between source and field point
% dxr ...   first derivatives of r
% vnor ...  outward normal
% E  ... modulus of elasticity
% ny ... Poissons ratio
% nstress  ... stress state indicator
% (1= plane strain, 2= plane stress, 3= 3D)
%
% OUTPUT:
% RK   ... derived Kernel
%-----------------------
G= E/(2.0*(1+ny));
ndim=2; nstr=3;
if(nstress == 1)
 c3= 1.0-2.0*ny; c4= 2; c6=4;c=1/(4*pi*(1-ny)*r^2);
elseif(nstress == 2)
 c3= (1.0-ny)/(1+ny); c4= 2; c6=4; c=(1+ny)/(4*pi*r^2);
```

```
elseif(nstress == 3)
 c3= 1.0-2*ny; c4= 3; c6=5; c7= 1- 4*ny;
 c=c=1/(8*pi*(1-ny)*r^3); ndim= 3; nstr=6;
endif
%  Voight notation
if( nstress < 3)
 I(1:3)= [1 2 1]; J(1:3)= [1 2 2];
else
 I(1:6)= [1 2 3 1 2 1]; J(1:6)= [1 2 3 2 3 3];
endif
for k=1:ndim
 for ns=1:nstr
  i=I(ns); j=J(ns);
  RKbar(ns,k)= Rbarijk(i,j,k,c3,c4,c6,ny,dxr,vnor);
 end
end
RKbar= RKbar*c
endfunction;
```

4 DERIVED SOLUTION FOR DISPLACEMENT S(x,y)

```
function S= Sijk(i,j,k,c3,c4,dxr)
% -------------------------
% Computes nonsingular part of derived Kernel
% for internal stress solution (to be multiplied with t)
%
% INPUT:
% i,j,k  ...  indices
% c3,c4  ... constants
% dxr ... first derivatives of r
%
% OUTPUT:
% S   ... derived Kernel
%-----------------------
S=0;
if(i == k) S=S+c3*dxr(j); endif
if(j == k) S=S+c3*dxr(i); endif
if(i == j) S=S-c3*dxr(k); endif
S=S+c4*dxr(i)*dxr(j)*dxr(k);
endfunction;

function SK  = SKernel(r,dxr,E,ny,nstress)
% -------------------------
% Computes derived Kernel
% for internal stress solution (to be multiplied with t)
```

```
%
% INPUT:
% r ...  distance between source and field point
% dxr ...   x/r,y/r
% E  ... modulus of elasticity
% ny ... Poissons ratio
% nstress  ... stress state indicator
% (1= plane strain, 2= plane stress, 3= 3D)
%
% OUTPUT:
% SK   ... derived Kernel
%----------------------
G= E/(2.0*(1+ny));
ndim=2; nstr=3;
if(nstress == 1)
 c= 1.0/(8.0*pi*G*(1.0-ny)*r); c3= 1.0-2.0*ny; c4= 2;
elseif(nstress == 2)
 c= (1.0+ny)/(8.0*pi*G*r); c3= (1.0-ny)/(1+ny); c4= 2;
elseif(nstress == 3)
 c= 1.0/(16.0*pi*G*(1.0-ny)*r^2); c3= 1.0-2*ny; c4= 3;
 ndim= 3; nstr=6;
endif
% Voight notation
if( nstress < 3)
 I(1:3)= [1 2 1];
 J(1:3)= [1 2 2];
else
 I(1:6)= [1 2 3 1 2 1];
 J(1:6)= [1 2 3 2 3 3];
endif
for k=1:ndim
 for ns=1:nstr
  i=I(ns); j=J(ns);
  SK(ns,k)= Sijk(i,j,k,c3,c4,dxr);
 end
end
SK= SK*c
endfunction;
```

5 DERIVED SOLUTION FOR TRACTION R(x,y)

```
function R= Rijk(i,j,k,c3,c4,c6,c7,ny,dxr,vnor)
% -------------------------
% Computes nonsingular part of derived Kernel
% for internal stress solution (to be multiplied with u)
```

```
%
% INPUT:
% i,j,k  ...  indices
% c3,c4,c6,c7,ny  ... constants
% dxr ...  first derivatives of r
% vnor  ... outward normal
%
% OUTPUT:
% R   ... derived Kernel
%-----------------------
cost= vecdotp(vnor,dxr); mult= c4*cost;
R=0;
if(i == j) R= R + c3*dxr(k); endif
if(i == k) R= R + ny*dxr(j); endif
if(j == k) R= R + ny*dxr(i); endif
R= R - c6*dxr(i)*dxr(j)*dxr(k);
R= mult*R;
if(i == k) R=R + c3*vnor(j); endif
if(j == k) R=R + c3*vnor(i); endif
R= R + c3*c4*dxr(i)*dxr(j)*vnor(k);
if(i == j) R=R - c7*vnor(k); endif
R=R + c4*ny*(vnor(j)*dxr(i)*dxr(k) + vnor(i)*dxr(j)*dxr(k));
endfunction;

function RK  = RKernel(r,dxr,vnor,E,ny,nstress)
% -------------------------
% Computes derived Kernel
% for internal stress solution (to be multiplied with u)
%
% INPUT:
% r ...  distance between source and field point
% dxr ...   x/r,y/r
% vnor ... outward normal
% E  ... modulus of elasticity
% ny ... Poissons ratio
% nstress  ... stress state indicator
% (1= plane strain, 2= plane stress, 3= 3D)
%
% OUTPUT:
% RK   ... derived Kernel
%-----------------------
G= E/(2.0*(1+ny));
ndim=2; nstr=3;
if(nstress == 1)
 c3= 1.0-2.0*ny; c4= 2; c6=4; c7= 1- 4*ny;c=G/(2*pi*(1-ny)*r^2);
elseif(nstress == 2)
 c3= (1.0-ny)/(1+ny); c4= 2; c6=4; c7= (1- 3*ny)/(1+ny);
 c=(1+ny)*G/(2*pi*r^2);
elseif(nstress == 3)
```

```
 c3= 1.0-2*ny; c4= 3; c6=5; c7= 1- 4*ny;
 c=G/(4*pi*(1-ny)*r^3); ndim= 3; nstr=6;
endif
% Voight notation
if( nstress < 3)
 I(1:3)= [1 2 1];
 J(1:3)= [1 2 2];
else
 I(1:6)= [1 2 3 1 2 1];
 J(1:6)= [1 2 3 2 3 3];
endif
for k=1:ndim
 for ns=1:nstr
  i=I(ns); j=J(ns);
  RK(ns,k)= Rijk(i,j,k,c3,c4,c6,c7,ny,dxr,vnor);
 end
end
RK= RK*c
endfunction;
```

6 DERIVATIVES OF KERNEL S(x,y)

```
function S= Sijkder(i,j,k,m,c3,c4,dxr,dxr2)
% --------------------------
% Computes nonsingular part of derivatives of S Kernel
% in tensor notation
%
% INPUT:
% i,j,k,m  ...  indices
% c3,c4  ... constants
% dxr ... first derivatives of r
% dxr2 ...  second derivatives of r
%
% OUTPUT:
% S   ... derivatives of Kernel
%-----------------------
S=0;
if(i == k) S=S+c3*dxr2(j,m); endif
if(j == k) S=S+c3*dxr2(i,m); endif
if(i == j) S=S-c3*dxr2(k,m); endif
S=S+c4*(dxr2(i,m)*dxr(j)*dxr(k) + dxr2(j,m)*dxr(i)*dxr(k)
    + dxr2(k,m)*dxr(i)*dxr(j));
endfunction;

function SKder  = SKernelder(r,dxr,dxr2,E,ny,nstress)
% --------------------------
```

```
% Computes derivatives of Kernel S
%
% INPUT:
% r ...   distance between source and field point
% dxr ...    first derivatives of r
% dxr2 ... second derivatives of r
% E  ... modulus of elasticity
% ny ... Poissons ratio
% nstress  ... stress state indicator
% (1= plane strain, 2= plane stress, 3= 3D)
%
% OUTPUT:
% SKder   ... derivatives of Kernel
%-----------------------
ndim=2; nstr=3;
if(nstress == 1)
 c2= 1.0/(4.0*pi*(1.0-ny));c3= 1.0-2.0*ny; c4= 2;c=c2/r;
elseif(nstress == 2)
 c2= (1.0-ny)/4.0*pi; c3= (1.0-ny)/(1+ny); c4= 2; c=c2/r;
elseif(nstress == 3)
 c2= (1.0-ny)/(8.0*pi*(1.0-ny)); c3= 1.0-2*ny; c4= 3;
 c=c2/r^2; ndim= 3; nstr=6;
endif
%  Voight notation
if( nstress < 3)
 I(1:3)= [1 2 1];
 J(1:3)= [1 2 2];
else
 I(1:6)= [1 2 3 1 2 1];
 J(1:6)= [1 2 3 2 3 3];
endif
for m=1:ndim
 for k=1:ndim
  for ns=1:nstr
   i=I(ns); j=J(ns);
   SK= Sijk(i,j,k,c3,c4,dxr)*dxr(m);
   SKder(ns,k,m)= -SK/r + Sijkder(i,j,k,m,c3,c4,dxr,dxr2)*c;
  end
 end
end
endfunction;
```

7 DERIVATIVES OF KERNEL R(x,y)

```
function R= Rijkder(i,j,k,m,c3,c4,c6,ny,dxr,dxr2,vnor)
% -------------------------
```

```
% Computes derivatives of Kernel R
% for internal stress solution (to be multiplied with u)
% in tensor notation
%
% INPUT:
% i,j,k,m  ... indices
% c3,c4,c6,c7,ny  ... constants
% dxr ... first derivatives of r
% dxr2 ... second derivatives of r
% vnor  ... outward normal
%
% OUTPUT:
% R    ... derived Kernel
%-----------------------
R=0; dim=length(dxr);
mult=0;
for n=1:dim
 mult= mult + dxr2(n,m)*vnor(n);
end
mult= mult*2;
if(i == j) R= R + c3*dxr(k); endif
if(i == k) R= R + ny*dxr(j); endif
if(j == k) R= R + ny*dxr(i); endif
R= R - c6*dxr(i)*dxr(j)*dxr(k);
R= mult*R;
mult=0;
for n=1:dim
 mult= mult + dxr(n)*vnor(n);
end
mult= mult*2;
R1=0;
if(i == j) R1= R1 + c3*dxr2(k,m); endif
if(i == k) R1= R1 + ny*dxr2(j,m); endif
if(j == k) R1= R1 + ny*dxr2(i,m); endif
R1= R1 - c6*(dxr2(i,m)*dxr(j)*dxr(k) + dxr2(j,m)*dxr(i)
    *dxr(k) + dxr2(k,m)*dxr(i)*dxr(j));
R1= R1*mult;
R2= vnor(i)*(dxr2(j,m)*dxr(k) + dxr2(k,m)*dxr(j)) + vnor(j)*
    (dxr2(i,m)*dxr(k) + dxr2(k,m)*dxr(i));
R2= R2*2*ny;
R2= R2 + c3*2*vnor(k)*(dxr2(i,m)*dxr(j)+dxr2(j,m)*dxr(i));
R= R + R1 + R2;
endfunction;

function RKder  = RKernelder(r,dxr,dxr2,vnor,E,ny,nstress)
% -------------------------
% Computes derivatives of Kernel R
```

```
%
% INPUT:
% r ... distance between source and field point
% dxr ...   first derivatives of r
% dxr2 ... second derivatives of r
% vnor ... outward normal
% E  ... modulus of elasticity
% ny ... Poissons ratio
% nstress  ... stress state indicator
% (1= plane strain, 2= plane stress, 3= 3D)
%
% OUTPUT:
% RK   ... derived Kernel
%-----------------------
G= E/(2.0*(1+ny));
ndim=2; nstr=3;c=G/(2*pi*(1-ny)*r^2);
if(nstress == 1)
 c3= 1.0-2.0*ny; c4= 2; c6=4; c7= 1- 4*ny;
elseif(nstress == 2)
 c3= (1.0-ny)/(1+ny); c4= 2; c6=4; c7= (1- 3*ny)/(1+ny);
elseif(nstress == 3)
 c3= 1.0-2*ny; c4= 3; c6=5; c7= 1- 4*ny;
 c=G/(4*pi*(1-ny)*r^3); ndim= 3; nstr=6;
endif
% Voight notation
if( nstress < 3)
 I(1:3)= [1 2 1];
 J(1:3)= [1 2 2];
else
 I(1:6)= [1 2 3 1 2 1];
 J(1:6)= [1 2 3 2 3 3];
endif
for m=1:ndim
 for k=1:ndim
  for ns=1:nstr
   i=I(ns); j=J(ns);
   RK= Rijk(i,j,k,c3,c4,c6,c7,ny,dxr,vnor);
   RKder(ns,k,m)= -RK*2/r*dxr(m) +
       Rijkder(i,j,k,m,c3,c4,c6,ny,dxr,dxr2,vnor)*c;
  end
 end
end
endfunction;
```

BIBLIOGRAPHY

[1]   G. Beer, I. Smith, and C. Duenser. *The Boundary Element Method with Programming.* Springer-Verlag, Wien, 2008.

[2]   M. Noronha, A.S. Müller, and A.M.B. Pereira. A novel pure-BEM approach for post-processing and non-linear analysis. In *Proceedings of MacMat2005, Louisiana, USA,* 2005.

Subject index

Note: Page numbers followed by *n* refer to footnotes

assembly, 132, 147, 168–169, 211–213, 218, 225

Bernoulli beam theory, 117, 153
Bernstein polynomials, 22–24, 29
Bézier curves, 24*n*5
body forces, 137, 139, 183, 187–188, 260, 264–266, 268, 270–275, 277, 286–287, 289–290, 295
B-splines, 24–31, 38–39, 41–42, 47, 50, 106*n*3, 121–126, 128, 132–135, 140, 150, 163–164, 168, 171–172, 175

collocation, 196–197, 199
collocation points, 202–209, 211–213, 218–220, 222, 224–225, 227, 229, 235–239, 243–245, 249, 252–254, 256–257, 271, 280, 284–286, 289–292, 294
consistent mass matrix, 300
control points, 23–24, 31, 40–41, 45–47, 57, 59–62, 66, 69–70, 72–74, 77, 79–83, 85, 87, 90, 97, 101–102, 104, 106–109, 111–112, 121, 135, 142, 146, 149, 158–160, 165–166, 206–207, 216–217, 220, 227, 229, 232, 235–236, 244, 246, 252–254, 256, 267, 269, 284–285, 290, 292, 294
control polygon, 24, 67, 97

Dirac Delta, 176, 183, 187–188
Dirichlet conditions, 139, 139*n*6, 140–142, 158–159, 169, 210–212, 217, 218*n*1, 226

Eigenfrequencies, 300
Eigenperiod, 302, 309

elasto-plastic, 265, 276, 279–280, 283–285, 287, 295
Element stiffness matrix, 138
Element, 131–132
entity, 98*n*2, 100–102

finite difference method, 3–4, 301–302
fundamental solutions, 4, 6, 12, 175–179, 183–186, 199, 214, 216, 251, 263, 265, 271–272, 281, 311–322

Galerkin method, 196, 199
Gauss integration, 124–125, 205, 268, 270, 273, 291
generatrix, 73–74, 77–81, 100

h-refinement, 128–132, 143

inclusions, 260, 263–266, 271–275, 289, 295
initial stress, 214, 263–264
isoparametric, 11–13, 140–141, 201–206, 232, 253, 258–259
isotropic, 176

Jacobian, 11, 57, 59, 63–65, 67–68, 70, 72, 77–78, 84–87, 89–90, 92–93, 112, 121–122, 124–126, 136, 138, 140, 155, 194, 205, 208–209, 214, 218, 222–224, 236–237, 243–246, 268, 270, 272, 285–286, 289–291
Jacobi matrix, 84, 147, 268, 270, 273

Kernel, 175, 184–187, 190–191, 194–195, 201, 203–205, 214, 216, 224, 236–239, 243, 246, 268, 270, 286–287, 289–291, 293, 312–322
Kirchhoff-Love theory, 153

T - #0498 - 071024 - C344 - 246/174/15 - PB - 9780367783433 - Gloss Lamination